T0261317

PLATO AND THE INVENTION OF LIFE

Plato and the Invention of Life

MICHAEL NAAS

FORDHAM UNIVERSITY PRESS

New York 2018

Copyright © 2018 Fordham University Press

All rights reserved. No part of this publication may be reproduced, stored in a retrieval system, or transmitted in any form or by any means—electronic, mechanical, photocopy, recording, or any other—except for brief quotations in printed reviews, without the prior permission of the publisher.

Fordham University Press has no responsibility for the persistence or accuracy of URLs for external or third-party Internet websites referred to in this publication and does not guarantee that any content on such websites is, or will remain, accurate or appropriate.

Fordham University Press also publishes its books in a variety of electronic formats. Some content that appears in print may not be available in electronic books.

Visit us online at www.fordhampress.com.

Library of Congress Cataloging-in-Publication Data available online at https://catalog.loc.gov.

Printed in the United States of America

20 19 18   5 4 3 2 1

First edition

for DFK

CONTENTS

Philosophy's Gigantomachia over Life and Being

In a word, *zōē* is a word for being . . .

—DAVID FARRELL KRELL[1]

"Being"—we have no idea of it apart from the idea of "living."—How can anything dead "be"?

—FRIEDRICH NIETZSCHE[2]

In his 1997 book devoted to the work of his long-time friend Hélène Cixous, a book whose title, *H. C. for Life*, at once names its theme and pronounces its dedication, Jacques Derrida writes the following about the question of life in the Western philosophical tradition:

> in the philosophical gigantomachia [that runs] from Plato to Descartes, from Nietzsche to Husserl, Bergson, and Heidegger, among others, the only big question whose stakes remain undecided would be to know whether it is necessary to think being [*l'être*] before life [*la vie*], beings [*l'étant*] before the living [*le vivant*], or the reverse.[3]

It is a rather sweeping, ambitious claim, to be sure. Derrida is asserting here that in the long philosophical tradition that runs from Plato to Heidegger, the only big question, *la seule grande question*, he says, that has yet to be settled, the only question whose stakes remain undecided, is the question of the relationship between being and life. Derrida will go on in this book, using especially the literary works of Cixous, interestingly, to provide a radical rethinking of this relationship, a rethinking of power or *puissance* in relationship to both being and life. It is in this way that Derrida not only names the gigantomachia but himself joins into the fray, trying to displace and reinscribe, it could be shown, the terms of both an ontological and a vitalist account of this relationship.

Now in speaking here of a gigantomachia, Derrida is, of course, alluding to the famous passage from Plato's *Sophist* where the Eleatic Stranger says that, when it comes to the nature of being or not-being, "there seems to be a battle like that of the gods and the giants [*gigantomachia*] going on among them, because of their disagreement about existence" (246a).[4] As the Stranger will go on to tell the tale, the battle pits those whose weapons are "derived from the invisible world alone," in short, the idealists or the friends of the forms, against those who "drag down everything from heaven and the invisible to earth," namely, the atomists or the materialists more generally (*Sophist* 246a–b).[5] It is a gigantomachia that was well underway when Plato wrote the *Sophist* and it is one that would continue to rage, it seems, all the way up to Heidegger and beyond. It is no coincidence, then, that Heidegger himself would refer to this well-known passage on the gigantomachia of philosophy in the very first paragraph of the Introduction to *Being and Time*, just half a page after his opening epigraph to the entire work, which is also from the *Sophist* and is today probably just as well known as that other passage: "it is clear that you have known all along [what you wish to designate when you say 'being'], whereas we formerly thought we knew, but are now perplexed" (*Sophist* 244a).

In its philosophical context, then, or at least in its Platonic inscription, the term *gigantomachia* designates the battle between those who begin with what is empirical or material, visible or touchable—that is, the giants—and those who believe that everything is derived from some invisible, ideal realm, the Olympians. The struggle is over the nature of what is, the nature of existence, *ousia*, or being, *to on*, and, by extension, non-being. But it is not, notice, as Derrida suggests in *H. C. for Life*, over the relationship between being and life. The *name* gigantomachia may thus go back to Plato but not necessarily the *thing* as Derrida has characterized it. One might therefore argue that Derrida has projected onto Plato a theme or a question that is central, in one way or another, for the Pre-Socratics, for Aristotle, or for the neo-Platonist tradition, beginning with Plotinus, but not necessarily for Plato himself, who is, to be sure, always concerned with what it means to lead a good or virtuous life but who does not seem so concerned with the question of life itself or the relationship between being and life.

And yet references to life, words and ideas related to life, can be found everywhere throughout the dialogues. In the *Sophist* itself, for example, at the precise point where the gigantomachia between the idealists and the materialists over the nature of being is invoked, it is a reference to life that is chosen to draw out and so distinguish the two parties. In order to see who will affirm the existence of an invisible as well as a visible realm, the Stranger says they should ask each party "whether they say there is such a thing as a mortal animal [*thnēton zōon*]," understood as "a body with a soul in it [*sōma empsychon*]," and whether the two sides give "to soul a place among things which exist" (*Sophist* 246e).[6] In order to distinguish the ideal from the material, therefore, Stranger introduces, at least implicitly through this reference to a *zōon*, the question of life. He could have asked any number of other questions; for instance, he could have asked both sides (following the *Euthyphro*) whether one does not need a form of piety in order to call some particular person or action pious, or he could have asked (following Book 10 of the *Republic*) whether a craftsman does not need an idea or model of a table before his mind in order to make an actual table. But the Stranger chose instead the example of a *zōon*. He did so, it seems, because that example presented the best, clearest, quickest, least contentious, and perhaps liveliest way of getting both sides to admit that there must be both a visible and an invisible realm, a realm of bodies and a realm of bodiless things such as the soul, or at least of qualities such as wisdom and justice that are traditionally attributed to the soul. The Stranger's explicit aim is surely not to introduce the notion of life into the discussion by means of this example. He does not, after all, go on to ask about the nature of this life, whether it resides in the soul, in the body, or only in the conjunction of the two. He is merely trying to distinguish a realm of invisible, ideal things, from material ones, and opposing soul and body seemed like the best or quickest way of achieving that end. And yet the words *zōon* and *empsychon* manage nonetheless to introduce the question of life into the dialogue and, once introduced, the question never simply disappears.

This, I will argue, is precisely how Plato introduces the notion of life into most of his dialogues, *using* rather than *mentioning* it in one dialogue after another, particularly in early and middle dialogues, but then mentioning it as well as using it, *zōē* as well as *bios*, in later dialogues. While the question of life is thus never the explicit focus of any Platonic dialogue, while it appears to be of minor importance next to the great Platonic questions regarding justice or the Good, the existence of the forms or the immortality of the soul, it is, we will see, an absolutely central, structuring question in all of Plato's dialogues. If there is, therefore, no Platonic dialogue that has or that really deserves the subtitle "On Life" (*peri biou* or *peri zōēs*), no dialogue that poses the famous *ti esti* or "what is" question with regard to life, I will argue here that almost everything in Plato's dialogues can and should be read through this question. From the question of how best to live a uniquely human life to the question of what distinguishes human life from other kinds of life, whether that of plants, other animals, or the gods, almost all of Plato's ethical, political, and even epistemological questions revolve around the theme or question of life.

In this work, however, it is especially the relationship between life and ontology that will be of most concern, the way in which Plato's dialogues tend to characterize being itself, albeit almost never directly and almost always just in passing, in terms of life or the living. It is this association of being with life that will allow Plato to oppose life not just to death but to everything that is typically opposed to being, everything from becoming and phenomenality to corporeality and animality, everything, in short, that we commonly *call* life. We will thus see that the life that Plato privileges in several of his most important later dialogues is, in the end, not bare life (though there is, as we will see, such a notion in Plato) but not a good life either, a good human life in excess of that bare life, but something like *real life* or *life itself*—that is, a life beyond or in excess of what we call life. By thinking being (whether in the form of the Forms or the immortal soul) in terms of life, Plato is ultimately able to discover (or, as I will argue, to *invent*) this notion of *life itself*—that is, a *real* or *true* life that would be opposed to all merely *biological* or *animal* life, a form of life, then, that would be more valuable than everything we call life and every life that can actually be lived.[7] As we will see, it is Plato's initial separation of bare life from human life, mere life from a truly worthy human life, that will facilitate this transformation and elevation of bare life into *real life*. In other words, it will be the *line that is drawn* between bare life and human life that will allow Plato to reinscribe the first of these two terms into a new, higher register—that is, into the register of *life itself.*

We will come to see in what follows that the gigantomachia that Derrida speaks of in *H. C. for Life* as taking place between Plato and others—all the way up to Heidegger and beyond—is already raging *in* or *within* Plato himself. The gigantomachia is more or less everywhere, just below the surface, and it is responsible for the division not only between bare life, the good life, and life itself, but between so many time-honored terms in the Platonic corpus, beginning with soul and body, being and becoming, the invisible and the visible, the ideal and the material, the one and the many, in other words, the entire matrix we call Platonism. This question of life in the Platonic dialogues will not only illuminate the structural relationship between all these terms and distinctions in Plato but will help to explain the enormous power and authority that Plato's thought will have exercised, for good or ill, over our entire philosophical and religious tradition. It will help explain, to cite an example to which I will return at the very end of this work, the Neo-Platonists' very explicit identification of Being, Logos, and the One with Life itself, a decisive moment for Western philosophy and theology.

To broach this question or these questions of life in Plato—this "swarm" of questions, to use a figure from the *Republic* (450b)—I will range far and wide throughout the Platonic corpus, reading passages from many dialogues, from *Alcibiades I*, *Protagoras*, *Gorgias*, and *Phaedrus*, to *Theaetetus*, *Sophist*, *Philebus*, *Timaeus*, and *Laws*, to name just a few. But much of this work will be devoted to the reading of a single dialogue, the *Statesman*, and, especially, the famous myth of the two ages where Plato seems to sketch out not only two conceptions of political rule but two different valences, values, or even types of life. A focus on life in the *Statesman* will thus be my starting point for asking the question of life in Plato more generally, the question of what distinguishes different forms of life from one another and the question of whether there is not some complicity between the ontological question "what is life?" and *life itself*.

The choice of the *Statesman* is, admittedly, not an obvious one. While no Platonic dialogue, as I said, asks the question "What is life?" (*ti esti ho bios* or *ti esti hē zōē*) in an explicit or sustained fashion, other dialogues would seem better suited for posing this question. There is the *Philebus*, for example, which asks not just, as in the *Apology*, whether the examined life is the only life worth living but whether a life of pleasure is preferable to a life of the mind, or whether a mixture of the two is preferable to the pure version of either. There is also, of course, the *Timaeus*, which offers an account of the origin of both the universe, itself a living being, and all the other living beings or creatures within it. And then there is the *Phaedo*, where, in the course of an inquiry into the nature and immortality of the soul, Socrates seems to posit a form of life—a life form—that is responsible for making everything it takes hold of or everything that participates in it alive. These are, as we will see, all important dialogues for the question of life in Plato and I will have occasion to look in some detail at key moments in them all. But none of these dialogues offer us, I will try to argue, as powerful a point of entry into the question of life in Plato's dialogues as the *Statesman*.

The theme and vocabulary of life are woven throughout the entirety of the *Statesman*, even if this may not be evident at first glance. An implicit notion of life is in the background animating, as it were, all of Plato's arguments and distinctions, from the initial division to define the statesman to the myth of the two ages to the separation of two forms of measure to the distinction between unwritten and written law. It is this largely unspoken but nonetheless nearly omnipresent preoccupation with life that will motivate Plato to draw the line not just between life and non-life but between the human and the animal, soul and body, speech and writing, presence and absence, legitimacy and illegitimacy, the Age of Kronos and the Age of Zeus, an age of life and fecundity and an age of death and forgetting. It is, I will argue, the theme of life in all these different contexts, this common reference to life, that will allow me to think together the various themes and movements of what has often been considered to be a rather disjointed or even ill-fashioned dialogue. As I will show, none of the parts of this dialogue can be fully understood in relation to the whole, and none of the above-mentioned terms and oppositions can be understood in relation to one another, without reference to this underlying theme of life.

What follows is not, however, a line-by-line commentary on the *Statesman*, even if I do hope to provide, in the course of this work, a relatively full and coherent reading of this dialogue. The last two or three decades have seen the publication of several important commentaries on this—up until then—much neglected dialogue. I will consider several of these commentaries, but I will focus especially on three—those of Stanley Rosen, Mitchell Miller, and my colleague at DePaul, David White.[8] I will not dispute that the *Statesman* can and should be read as a dialogue that attempts to demonstrate that the true statesman is a philosopher, as Stanley Rosen argues, or that it is essentially a dialogue about philosophical pedagogy, as Mitchell Miller claims, or that it is a thoroughly aporetic dialogue that calls for a reading of other dialogues, and particularly the *Philebus*, on the nature of the Good, as David White maintains.[9] I will instead try to show that the *Statesman* is even more significantly a dialogue about Plato's conception of life and everything that goes along with it in his philosophy, beginning with statesmanship and pedagogy within human life and ending with the Good as the very source—perhaps the living source—of life itself.

Two more preliminary remarks, one related to the theme of this work and one to the methodology or approach it takes to its subject. This work focuses on the question of life in Plato because this question can illuminate a great deal—indeed almost everything, as I have suggested—about Plato's work. But I also focus on it because the most insistent question over the past two or three decades in contemporary European philosophy or what is known as Continental Philosophy has no doubt been the question of life: the question of the limits of life, the relationship between life and its others (death, the inorganic, the technological, the machinic), the question of biopower and the regulation of life, the relationship between various kinds of life (the human, the animal, the plant), even the question of "life itself" in recent discourses of so-called new materialism. In a recently published volume on *Philosophy in France Today*, Frédéric Worms documents this concern with life in contemporary French philosophy in its relationship to science, politics, metaphysics, and the history of philosophy, and he remarks on the fact, which he takes to be hardly a coincidence, that the final texts of Foucault, Deleuze, and Derrida were all concerned with the question or theme of life.[10] I thus hope to provide some contribution, through this reading of Plato, to a discussion about life in contemporary philosophy.

With regard to methodology, I will be working here essentially with the Platonic text, in its arguments, its themes, its dramatic context, and its letter. But no one, of course, ever approaches a text without some orientation or methodological presuppositions, some interpretative horizon.[11] My approach to the question of life in Plato's dialogues will thus be informed, in part, by the recent commentaries on the *Statesman* that I just mentioned, as well as by Jacques Derrida's "Plato's Pharmacy."[12] In that seminal essay of 1968, Derrida demonstrated, in effect, that Plato was able to develop—indeed, to *invent*—a conception of speech without or before writing only by means of what he calls the *supplement* of writing. I will deploy this same "logic of the supplement," as Derrida called it, in order to show how Plato was able to discover—indeed, to invent—a notion of life that would supposedly come without or before all death and, thus, without or before everything that is *called* life, an Age of Kronos that would precede and exceed every Age of Zeus. Just as Derrida in "Plato's Pharmacy" focuses on the myth regarding the nature of writing at the end of the *Phaedrus* as a way of entering into Plato's thinking about speech and writing more generally, so I begin with the *Statesman*, and particularly its myth of the two ages, in order to approach the question of life and death more generally in the Platonic corpus.

My reference to "Plato's Pharmacy" suggests yet another reason for focusing on the *Statesman* in this work. It, not the *Phaedrus*, is the dialogue with which Derrida actually begins his long, hundred-and-twenty page essay. After evoking in its opening lines the image of the text as a web (rather than as an organic being—that is, as a *zōon*), Derrida writes: "The example we shall propose of this will not, seeing that we are dealing with Plato, be the *Statesman*, which will have come to mind first, no doubt because of the paradigm of the weaver, and especially because of the paradigm of the paradigm, the example of the example—writing—which immediately precedes it. We will come back to that only after a long detour" ("PP" 65/74). Derrida will indeed return to the *Statesman* at the very end of "Plato's Pharmacy," but just for a line or two, as he looks at the notion of the *symplokē* in Plato's later dialogues ("PP" 165/191, 166n82/192n77). He could have said much more about the *Statesman*, however, insofar as it turns around many of the same themes as the *Phaedrus*, including the question of writing as a supplement and, as we will see, the question of *life*. *Plato and the Invention of Life* thus tries to fill out, as it were, some of the ideas that are only suggested or left in outline form in Derrida's 1968 essay. It offers a reading of the *Statesman* that is inspired by Derrida's analyses in "Plato's Pharmacy" but that also draws from Derrida's writing and thinking more generally.

For example, throughout his seminars and his writings, Derrida would often ask himself, as he was reading a text, the question of where a thinker or text is trying to *draw the line*. Where, for instance, he asks repeatedly throughout his final seminar *The Beast and the Sovereign*, is Heidegger trying to draw the line between man and the animal in his *Fundamental Concepts of Metaphysics*? Or, already back in "Plato's Pharmacy," where does Plato draw the line in the *Phaedrus* between two forms of repetition or two forms of memory, corresponding, perhaps, to two forms of life? The question posed in this work will also be, then, where, when, why, and how does

Plato draw the line between life and its others, the line between life and being, at the outset of this great gigantomachia, but then also the line between life and non-life, life and death, the living and the dead, human life and its others, though also, and this will be my most speculative moment, the line between a life beyond life—life itself—and what we *call* life? Where does Plato or where do Plato's dialogues draw the line between true life, between what is most real and most living in life, and this mortal life that always ends in death? Where do Plato's dialogues draw the line in each of these cases in order to define and to know? But then also, where do these ontological or epistemological lines become political or ethical—that is, where does Plato, as we say, *draw the line* not just in order to distinguish between two things, concepts, or categories, but in order to stake out a boundary or limit so as to affirm or deny, say yes or no, in order to value some things, like life and especially human life, and reject or devalue others, such as non-life or non-human life, though also, and this will be, again, my most speculative moment, the life of being as opposed to the non-life or the quasi-life of becoming? It will be my contention that we can learn a lot about Plato by asking with persistence and penetration the question of where the line is to be drawn, and then, finally, what it is that eludes or else makes possible any drawing of a line in the first place. While one might be tempted to call such a reading a *deconstruction* of Plato's text, I would simply prefer to call it an exercise in *drawing the line*, an exercise in asking why Plato thought he had to draw the line in such a way so as to include some things and exclude others, value some things and devalue others, elevate some things and reject others.

I say I get this practice or this discipline of drawing or following lines from Derrida, but it would be right to point out that it goes all the way back to Plato and has been an essential part of philosophy ever since. The method of *diairesis*, for example, which is so central to the *Statesman*, as we will see in Chapter 1, is nothing but a sophisticated practice of drawing lines. One of the most studied and celebrated figures in all of Plato's dialogues is also, of course, a line, a divided line, which establishes hierarchies between different levels of being, knowledge, clarity, goodness, and so on.[13] It is by means of this simple line that Plato gives us a stunningly clear and yet extremely sophisticated vision of his entire ontology, epistemology, aesthetics, and, insofar as this line is introduced in a section of the *Republic* devoted to the philosopher king, ethics and politics. Though we are not going to analyze in any detail that divided line in this work, we will inevitably catch a glimpse of it at almost every turn, especially as we try to think near the end of this work what it is that makes this line possible without itself having a place on the line—a certain conception not of the good life or of life itself but of the Good *as* Life. Every time Plato draws a line, I will argue, even the most banal and seemingly insignificant, we catch a glimpse of that which exceeds the line and all the distinctions that are made on the basis of it, the source of all difference and—as I will argue in conclusion— *all life.*

We will thus remain perfectly faithful to a central aspect of Plato's thought if we simply persist in asking the question of where we *do* or *should* draw the line between one term, concept, or idea and another. This is the case not only for already well-known and highly valued relations but for those that are less well-known and appear much less valued. As such, we will become better able to understand Plato's entire ontology, epistemology, aesthetics, ethics, politics, and so on by asking how and why Plato draws the line not just between the intelligible, invisible realm, on the one hand, and the sensible, visible one, on the other, or between being and becoming, the one and the many, but between life and its others, life and death, to be sure, but also spontaneity and automatic movement, living memory and mechanical recollection, legitimate heirs and bastard offspring, the organic and the inorganic, fertility and sterility, human life and animal life, animal life and plant life, to name just a few. In each of these lines we can find, if we look closely enough, all these other distinctions, all these other lines, from the most noble and traditional to what seem to be the most trivial or banal. It is Plato, well before Derrida, who showed us that every line is implicated in every other line. We will thus want to multiply these lines in order to see the relationships between them but then also, as I have suggested, in order to locate a notion of life that exceeds them, a notion of life that might itself account for this prodigious proliferation of differences, this quickening of differences and relations in the dialogues.

If I thus often argue in this work for a rather Platonic or Platonist reading of Plato, I will get there through what are some rather atypical Platonic or Platonist questions, and I will do so in order to arrive at something that escapes Platonic categories altogether. I will, of course, look at the arguments presented in Plato's dialogues, the distinctions he appears to affirm and the ideas he seems to endorse. But I will tend to focus even more on Plato's language, on his vocabulary, his rhetoric, his use of myth, on the *matrix* of terms he *uses* as much as *mentions*. It is in this way that I hope to expose both an undeniable Platonism in Plato's dialogues and that which contributes always to the undoing of that Platonism—right down to and including the category of *life*.

What follows, then, is an attempt to read large sections of the *Statesman*, in conjunction with parts of many other dialogues, on this question of *life* and the *line*. By focusing on the question of life, we will come to see that another logic is at work throughout Plato's *Statesman* and that this logic is essential to understanding not only the *Statesman* but Plato's work as a whole. The first six chapters of the work analyze in detail several sections of the dialogue with a view to this theme or question of life. In Chapter 1, I look at the attempt, in the opening pages of the *Statesman*, to discover the essence of the statesman through that method known as *diairesis*, which, as I suggested a moment ago, is nothing if not an exercise in learning how to draw the line or, as the Stranger would have us believe, learning how to cut along the lines that are already drawn in nature. In addition to asking about the place where Plato draws the line between life and its others, human life and its others, I will ask here about Plato's repeated use of animal metaphors to characterize the very method of philosophy—the dialogue as hunt, *diairesis* as animal sacrifice, Platonic forms as "natural species," and so on.

In Chapter 2, I focus on the myth of the two ages and the pivotal role played by the adjective *automatos* in the Stranger's telling of that myth. I argue that Plato's use of the term *automatos* to describe *both* the way in which the fruits of earth come up "of their own accord," "freely," or "spontaneously" during the Age of Kronos *and* the way the universe or cosmos, itself a living being, moves during the Age of Zeus, reveals a tension at the very heart of Plato's conception of life. By comparing Plato's use of this term in the *Statesman* to other uses in the dialogues (in *Theaetetus*, *Sophist*, *Protagoras*, and elsewhere), I argue that the term *automatos* must be understood as something like—to use Derrida's vocabulary—an *undecidable*, akin to the *pharmakon* of *Phaedrus*, that is, a fundamentally ambivalent notion around which a whole series of opposing terms (activity/passivity, inside/outside, natural movement/mechanical causality) revolves. I conclude by suggesting that this use of *automatos*, in conjunction with a somewhat unexpected use of the term *mimēsis* in the same myth, might indicate an unacknowledged Heraclitean influence on the *Statesman* and might help explain Plato's multiple references to Heraclitus in related dialogues (such as *Sophist, Theaetetus,* and *Cratylus*). This will allow me to ask whether a (perhaps Heraclitean) notion of *mimēsis*—another way of describing what I will call Plato's *political anamnesis*—is not what links the Age of Zeus to the Age of Kronos, and, perhaps, human *life* in the Age of Zeus to the *true life* of the Age of Kronos.

In the following chapter, I look at Michel Foucault's provocative but very selective reading of this same myth of the *Statesman* in his 1977–78 seminar *Security, Territory, Population.* Contra Foucault, who claims that Plato never adhered to the pastoral model of governing—that is, the model of the ruler as shepherd—and that he, in fact, definitively rejects this model in the *Statesman* in favor of the model of the ruler as weaver, I argue that a more complex relation between the models, as between the two ages, is required. Though Plato never puts it in exactly this way, the statesman in the Age of Zeus is indeed a weaver, but he is a weaver who must, through his weaving, *imitate* the shepherd in the Age of Kronos. Hence the model of the statesman as weaver—that is, the model of the statesman under-stood through an image of technical production within the city—must re-main, in a way that will have everything to do with the question of life, the statesman as shepherd in a time before technology and before the city.

In Chapter 4, "The Measure of Life and Logos," I look at Plato's dis-tinction between two forms of measure (relative measure and the measure of the mean) and the values of life they ultimately embody or represent. In the *Statesman,* Plato comes to see—and perhaps for the first time in his dialogues—that all production and, perhaps, all life actually depend on this measure of the mean, a notion that Plato had certainly used or deployed in dialogues before the *Statesman* but never commented on or analyzed as such. The chapter concludes by considering the implications of this mea-sure of the mean for logos or discourse, whether that discourse be merely rhetorical or, as in the *Statesman* itself, philosophical.

In Chapter 5, I return to the essay that I said inspired much of this work, Derrida's 1968 essay "Plato's Pharmacy," an essay that begins by focusing on the myth told by Socrates at the end of the *Phaedrus* about the inven-tion of writing. Derrida is able to show there how the relationship or op-position between speech and writing in the *Phaedrus* brings along with it Plato's entire philosophical matrix, from the relationships between unity and multiplicity, being and becoming, presence and absence, and memory and forgetting, to the oppositions between fecundity and sterility, legiti-macy and illegitimacy, and life and death. Cut off from the presence of a living speaker, written discourses present only a semblance of life and so threaten the real life that can be found only in living speech. In this chapter, I revisit Derrida's almost-half-century-old essay in order to ask about the values of life and living presence in Plato and in two rhetoricians/sophists with whom Plato seems to be in dialogue throughout the *Phaedrus* as well as the *Statesman*, Isocrates and Alcidamas.

While Isocrates and Alcidamas agree with Plato about the dangers of writing, they do so for very different reasons. For these two rhetoricians/sophists, written speeches are not only less true and less real than spoken ones but less efficacious, less able to take advantage of the critical moment (the *kairos*) in order to move or persuade an audience. For Plato, writing is dangerous to the extent that it is taken seriously as a genuine substitute for the spoken word or, worse, for that most interior of spoken discourses called "writing in the soul." I show that what is at issue between Alcidamas and Isocrates, on the one hand, and Plato, on the other, is not only the question of speech and writing, and the related questions of memory and repetition, but, once again, the question of life and the value of life. What is waged in the gigantomachia between the rhetoricians or sophists and Plato is a battle over two different values for life, life as force and the power to persuade, on the one hand, and life as truth or as the force of truth, on the other. What we see in all three thinkers is, as I will argue in conclusion, at once a fear of the power of the written word and an absolute fascination with it, a fear of this new technology called writing and a fascination with the transformation in the values of life that it at once threatened and promised.

In Chapter 6, I look at a large swath of the dialogue in order to ask, with "Plato's Pharmacy" in the background, about the Stranger's claim in the *Statesman* that law—and especially written law, inasmuch as writing is the essence of law—is at once inferior to rule without law and yet, in a world without divine rulers, absolutely necessary for human governance. This chapter returns to many of the insights from Chapter 2 on the myth of the two ages, since what that myth demonstrated was the desirability and yet impossibility of an age in which a truly divine being rules over human beings and the concomitant necessity of trying to *imitate* that age through laws. Once again, we will see that what is at issue in the relationship between the two ages, as well as in the relationship between a regime without law and a regime with it, are two different valences or valuations of life—the values of pure life, fecundity, spontaneity, and memory, on the one hand, and the values of death in life, forgetting in memory, and sterility in fecundity, on the other.

In the final chapter, I take many of the insights from the previous chapters in order to show, through a much wider reading of Plato's dialogues, how Plato attempts always to move from what is commonly called life, from a more biological conception of life, a life of the body or of the animal, to a spiritual life or a life of the soul—that is, from something like *bare* life to *real* life, from a merely phenomenal life to true life, from particular life-forms to an essence or form of life itself, the only life, in end, worthy of the name for Plato. I show here that, in Plato's early dialogues, this theme of life is almost always framed in exclusively ethical or political terms, that is, in terms of how best to live a uniquely human life or how best to organize life in the polis. In Plato's later dialogues, however, the notion of life begins to take on an ontological value in conformity with Plato's metaphysics. I thus look in this final chapter at several later dialogues where Plato begins to distinguish two different valences of life, human life in the polis (*bios*) as opposed to what Giorgio Agamben calls "bare life" (*zōē*), but also, and more importantly, human life as opposed to something like *real life*. As I will show, it is the initial distinction between human life and bare life that allows for this reinscription or transformation of bare life into something like *real life* or *life itself*, a transformation, as I will argue in conclusion, that is decisive not just for Plato but for the entire neo-Platonic and Christian tradition that takes its inspiration from him.

This move from life, from what is *called* life, to something like real life, true life, or life itself, this move to an essence or form of life, is, I will argue, one of the classical and most powerful moves or moments in what Derrida characterizes as the gigantomachia that runs from Plato to Heidegger. In his dialogues (and particularly the later ones) Plato begins to identify being more and more with life, with *zōē*, which will come to mean not bare life, not a uniquely human life, but life itself—that is, a life beyond or in excess of what *appears* to be life or what we *call* life and, thus, a life to which our finite human existences have but limited access. In the end, we see in Plato the beginnings of a form of life that, at the limit, must be thought apart from all living, mortal beings and, thus, apart from all death, a life that exceeds what we call life, a life that is the only life worth living even if it is a life that can never really be lived. In other words, for such a thinking of life, *la vraie vie est ailleurs*, real life is elsewhere.[14]

And yet, that is not Plato's final word on the matter of life—even if it will have been his most authoritative and influential. In the conclusion to this work, I suggest that there is in Plato, in addition to this undeniable tendency to think life in terms of life itself—that is, this tendency to distinguish two forms of life, one real and the other less real—another countervailing tendency, a less "Platonic" conception of life that is operating always in the margins of Plato's dialogues, a notion of life that needs to be thought along the lines of what Jacques Derrida began calling in an as-yet-unpublished seminar of 1975–76 *life death*.[15] I contend that what a Derridean reading of the *Statesman* and other dialogues on the question of life reveals is the necessity of thinking life otherwise—that is, life neither as bare, biological life, nor as real, spiritual life, as life itself, but as *life death*, as a life that must always be woven together with and thought always in relation to death. Just as the invisible can only be thought through and in relation to the visible, the soul through and in relation to the body, the one through and in relation to the many, so life must always be thought in relation to death—with no single or indivisible line drawn between them. *Life death*, I argue in conclusion, is what the philosophical and religious tradition of the West has had to forget or repress in order for something like life itself to emerge—that is, a life completely detached from any life that is actually lived. Such a reading will do nothing but reaffirm what I argued in earlier chapters, namely, that the myth of the two ages in the *Statesman* is not simply about some archaic past where the gods once ruled humans but the relationship—the line—between two forms of life, and the necessity of weaving these two forms together, life as what is simply opposed to death and life as what can never be thought without death.

If the question of the relationship between being and life is indeed central to the philosophical tradition, Plato included, then we will want to ask, finally, what the philosophical practice of drawing lines tell us about philosophy and life and about the relationship between them. For if, as I will argue in conclusion—and this will perhaps go beyond speculation properly speaking—a certain thinking of life is what makes all distinctions possible, then perhaps life must be thought, like *physis*, as that which loves to hide in the traces it leaves or the lines it draws. Life or *life death* would thus be inseparable from this very process or movement of drawing lines, whether that line be drawn between being and life, life and death, body and soul, or anything else. Where does Plato draw the line?—that is the question—but then also, what is the concept (if it is a concept) of life that would have allowed him to draw such a line or, rather, such lines, since as soon as there is a single line there are already many?

Since such a thinking of life never becomes an explicit object of investigation or sustained meditation in any of Plato's dialogues, we will have to develop it largely on our own, asking not just the *ti esti* question with regard to life but questions of another kind. For one of the hypotheses of this work will be that life, or *life death*, is what eludes the *ti esti* question, *life death* as the source of life, as the life-force of life, *life death* as the *symplokē* that at once distinguishes life from death, being from becoming, and withdraws from these distinctions, *life death* as what is always *epekeina tēs ousias*—that is, always beyond or in withdrawal from everything to which it gives rise, making it impossible to say, no matter how good it is and how clearly we think we see it, now *this* is the life.

CHAPTER I

The Lifelines of the *Statesman*

Où passe la ligne de partage entre l'événement d'un énoncé
inaugural, une citation, une paraphrase, un commentaire,
une traduction, une lecture, une interprétation?

—JACQUES DERRIDA[1]

Opening Lines

I begin my reading of the *Statesman* by briefly recalling the characters as
well as the dramatic time and setting of the dialogue (257a–258b). Present
for the dialogue is, first of all, Socrates, the Socrates we all know, the son
of Sophroniscus, from the deme of Alopece, the Socrates of *Euthyphro*,
Apology, *Crito*, *Phaedo*, and so many other early and middle dialogues, the
Socrates who, let me assert without justifying, is never simply the mouthpiece for Plato but, in most of these dialogues, surely represents the philosophical ethos and many of the views of their author. Also present in the
dialogue is, curiously, another Socrates, a younger one, he, too, from Athens, a character who therefore shares a name with Socrates and so, at least
in name, might be taken for or confused with him. In a dialogue in which
the question of the relationship or the fit between the name and the thing
is so central, a dialogue whose central myth depicts humans growing
younger and younger, it is surely not insignificant that it is Young Socrates

who will bear most of the responsibility for responding to the questions of the one who really leads the dialogue and has the most to say in it.[2] That is, of course, not Socrates, Socrates the Elder, as we might call him, who, as in *Sophist* and *Timaeus*, participates only at the outset of the dialogue, but someone called simply the Stranger (the *xenos*).

The dialogue is thus conducted in large part by a Stranger visiting Athens from Elea, in southern Italy, where he would have been part of the Eleatic school founded by Parmenides. While the Stranger's identification with Parmenides is much more in evidence in the *Sophist*, where he famously argues against his "father" Parmenides that non-being in a certain sense *is*, we will still want to ask throughout what this association with Parmenides signifies for the *Statesman* and why Plato gives such pride of place not to Socrates but to the Stranger in this important dialogue.[3] Again without justifying this for the moment, let me simply assert that if the Eleatic Stranger is no more the simple mouthpiece for Plato than Socrates ever is, it is hard to avoid the conclusion that some of his views would have been compatible with those of the author of the dialogue.[4] For if there is irony in the dialogue, and there surely is, and if there is skepticism with regard to certain methodologies, such as *diairesis*, and there surely is, it is hard not to think that the author of the *Statesman* would have endorsed at least some of what the Stranger argues. If, in addition, the Stranger ends up arguing things that would have sounded somewhat strange or incongruous coming out of Socrates's mouth, then we perhaps have a first reason for Plato's choice of the Stranger rather than Socrates to lead the dialogue.

Rounding out the dialogue, in addition to Socrates, the Young Socrates, and the Eleatic Stranger, is Theodorus, a mathematician from Cyrene (a colony in North Africa), also present in the *Sophist* and *Theaetetus*, and, finally, Theaetetus, a young man who, as we know from the *Theaetetus* (144d–e), bears a striking physical resemblance to Socrates.[5] We thus have one character whose name echoes Socrates's and another, Theaetetus, whose appearance resembles his, and whose hometown, Sunion, is enough to call Socrates to mind. For it was, as we know from the beginning of the *Crito*, from Cape Sunion, at the southern tip of Attica, that the ship returning to Athens from Delos was seen, portending the end to the holy period in Athens and thus the execution of Socrates. The mere presence of Theaetetus of Sunion in the dialogue is thus enough to evoke the entire drama surrounding "the last days of Socrates."

It is, however, not just the setting and characters of the dialogue that evoke Socrates's trial and execution but its dramatic time. Plato has Socrates say at the end of the *Theaetetus* that he must go off "to the Porch of the King, to answer the suit which Meletus has brought against [him]" (*Theaetetus* 210d). Socrates would have thus gone off to answer the indictment at the stoa of the King Archon and he would have encountered on his way, as we know, Euthyphro, whom he will engage in yet another, much less lengthy dialogue on the nature of piety. But before taking his leave at the end of the *Theaetetus*, Socrates suggests to Theodorus, who has been part of the dialogue, that they meet back up the following morning in the gymnasium for more conversation. Theodorus will keep his promise of meeting the next day, after Socrates will have conversed—according to the dramatic chronology—with Euthyphro and answered the indictment of Meletus against him. But when Socrates, Theodorus, and Theaetetus, the principal participants of the *Theaetetus*, meet the next day, Theodorus brings along with him the Eleatic Stranger, who will eventually lead a dialogue to define the sophist and then, later in the day, the statesman and, perhaps— though this never happens, or at least there is no such dialogue, though it seems to have been on the program—the philosopher.[6]

Like the *Theaetetus* and *Sophist*, then, the dramatic setting for the *Statesman* is a gymnasium and its dramatic date is 399 BCE. It immediately follows the *Sophist*, which follows by a day the *Euthyphro*, which takes place after but on the same day as the *Theaetetus*, and it would be followed, some days or weeks later, by *Apology*, *Crito*, and *Phaedo*. If everything from the style, theme, number of characters (just two), and the centrality of Socrates suggest that the *Euthyphro* was written long before *Theaetetus*, *Sophist*, and *Statesman*, Plato decided to wrap these latter three dialogues around the former, thus giving us four dialogues in a two-day dramatic stretch not long before the trial and death of Socrates. If, as I will argue in what follows, the *Statesman* needs to be rethought through the question of life, it will not be insignificant that it develops in an atmosphere or in the context of Socrates's impending death—that is, his impending absence and withdrawal.

The *Statesman* begins with Socrates thanking Theodorus for having introduced him the previous day, at the outset of the *Theaetetus*, to Theaetetus, and, earlier on the same day, to the Stranger, who went on in the *Sophist* to define the nature of the sophist (257a). When Theodorus says that Socrates will be three times more grateful once they have defined the statesman and the philosopher, Socrates rebukes Theodorus for using such a mathematical proportion, insinuating that there is no common measure between these three, the philosopher being on a different scale of value, it would seem, than the sophist or the statesman as they are commonly understood (257a–b).[7] One of the central questions of the dialogue will turn out to be whether or in what way the philosopher is indeed on a different scale of value and, in parallel fashion, whether the best form of government—the seventh form—appears on a different or the same scale of value as the other six that imitate it, but then whether, even more provocatively, a certain notion of being and of life, of life itself, of pure life, appears on a different or the same scale as various modes of becoming or various forms of what we *call* life.[8]

Having resolved to define the statesman after the sophist, the Stranger from Elea suggests using the Young Socrates rather than Theaetetus as his interlocutor so as to give the latter a rest (257c). It is here that Socrates remarks, in what will be his final words of the dialogue, that both Theaetetus and the Young Socrates are in some sense "related" to him, akin to him, insofar as one *looks* like him while the other *shares his name*.[9] Because one should, he says, "get acquainted with our relatives by debating with them," Socrates says he would be delighted to hear the Younger Socrates examined, having himself examined Theaetetus on the previous day. From this point on, it will be the Stranger who guides and orients the dialogue, questioning and leading the Young Socrates in a way that is not wholly unlike the way Socrates leads the young Theaetetus in the *Theaetetus*.[10]

Diairesis *as the Art of Drawing the Line*

The Stranger begins by taking up yet again, just as he had done in the *Sophist*, the method of *diairesis* or division, in short, the method of *drawing lines* (258b–267b). The word *diairesis*, from the verb *diaireō*, means precisely to distinguish, to divide, to draw lines. Though the Stranger himself will ultimately recognize the shortcomings or limitations of this method, it does not appear that the Stranger—or Plato, for that matter—is using it simply to criticize, mock, or reject it.[11] It is unlikely that this method would be used not only here but in the *Sophist* and the *Phaedrus* (265e) if Plato considered it without value. Moreover, if *diairesis* involves, at its most general level, trying to draw lines between concepts or ideas in order to highlight the relations and differences between them, then it could be said that all philosophy, including and especially dialectics, relies upon this practice, whether explicitly or implicitly. The use of *diairesis* at the beginning of the *Statesman* will therefore reveal as much about the object of the dialogue, namely, the statesman and statesmanship, as about philosophical method more generally.[12] It is no coincidence that the question of *method* will emerge so often in these early pages (see 258c, 262b, 263b, 265a), as if method—in this case the art or practice of drawing lines—were one of the central *themes* of the dialogue, as it is, of course, in so many of Plato's dialogues.

Once it has been decided that they will seek to define the statesman by means of *diairesis*, the next most important "decision" concerns the starting point for the division. This happens with little notice or fanfare when the Eleatic Stranger gets the Younger Socrates to affirm that the statesman, the *politikos*, belongs to the class of those who have a science, an *epistēmē* (258b). The object of the Stranger and Young Socrates's search will thus be the *politikos*, the statesman, the title of the dialogue itself, and his art or science, his *epistēmē*, would be *politikē*, statesmanship, an art or science whose end result or end product, at least as it is practiced by an originary statesman, would be the laying out of a *politeia*—that is, a constitution or a regime or, indeed, as this word is commonly translated in the title of one of Plato's most celebrated dialogues, a *republic*. Hence the entire *diairesis* is premised, it seems, upon a prior, unexpressed or unstated division between the arts or sciences, everything that might be called an *epistēmē*, and something else to which these arts and sciences would seem to be opposed. As we will see, everything will ultimately hinge on the fact that the statesman is credited here at the outset with having not a *tribē* (a knack) or *phronēsis* (practical wisdom) or even *sophia* (wisdom), but an *epistēmē* (a science), or, in some of the same passages, a *technē* (an art), these two words being used more or less interchangeably throughout the dialogue. Though we do not yet know the nature of the statesman's art or science, we do know that he has an art or a science and that it goes by the name of statesmanship.

Once this assumption or this assertion—this concession—has been made, the Stranger offers his first comment about the method of *diairesis*, arguing that they must divide the sciences in a way, or according to a path, that is very different from the one they followed when searching for the sophist (258b). The Stranger asks, "where shall we find the statesman's path? For we must find it, separate it from the rest, and imprint upon it the seal of a single class [*idean*]," that is, a single *idea* or single form. Notice the order here: They must first *find* this class, this *idea*, of the statesman, as if it already existed, even if virtually, in nature; they must then *separate* it off so that it can be more easily defined; and then they must label or *mark* it, put a sign, seal, or signet upon it, mark it as the unique idea that it is.

We go on to learn that this process of identifying and then marking a single idea is best achieved not by trying to draw off a single class from all the others at the outset, by trying to identify right from the start the statesman's art in relation to all the other arts that are not the statesman's, but by following a series of divisions. For even if one could immediately make out the statesman's class through a kind of immediate, intuitive apprehension, one would learn little about the exact nature of this art, the kind of science it involves, the sorts of objects with which it is concerned, and so on. To discover the greatest number of attributes of the object or category being sought or questioned, it is thus best to proceed, says the Stranger, by dividing as close as possible down the middle (261c), using distinctions ordered not in a series but in binary pairs, many of these being the very binaries we associate with Platonism: science/non-science, intellectual/practical, but, especially, as we will soon discover, soul/body, living/dead. As for later distinctions such as those between the horned and the un-horned, the feathered and the non-feathered, the biped and the four-legged, and so on, there is surely a good bit of humor involved, as Mitchell Miller among others has convincingly argued. But even these distinctions, however rudimentary and humorous, are not completely without utility and are not simply being criticized or mocked in the dialogue. Plato will have some fun with these distinctions at the end of the division, but this will hardly disqualify the method of *diairesis* itself. On the contrary, as we will also see, Plato appears to be keenly aware that philosophy as dialectics is always engaged in some form of *diairesis*—some form of line-drawing—whether it acknowledges it under this name or not.

Having already agreed that the statesman exercises a science, an *epistēmē*, the Stranger proceeds to divide the entire form or category of science into "pure sciences"—pure *technai*, as they are now called—sciences such as arithmetic that have no regard for practical application, sciences that merely furnish knowledge, and those such as carpentry and handicraft that aim at application and at producing objects or bodies that did not previously exist. There are, in short, two classes or forms, two *eidē*, of science, the intellectual (*gnōstikē*) and the practical (*praktikē*), and the question becomes to which of these two forms statesmanship belongs (258d).

Before answering this question, the Stranger gets the Young Socrates to make another concession, one that is crucial not only for the dialogue but, it would seem, for Plato's political theory in general. The Stranger argues that the science of statesmanship is the same whether it is exercised by a statesman, a king, or a householder—that is, regardless of the domain in which it is being exercised, regardless of the scope of one's rule and, perhaps most importantly, regardless of the *name* given to the person exercising it. To explain his point, the Stranger appeals to a distinction between the private person or the layperson and the professional (259a). If someone in the capacity of a layperson, an *idiōtēs*, is able to advise a public physician, it must be because he shares in the same art as the physician—namely, the art of medicine. Similarly, if a private man is able to advise a king, it must be because he shares in the same art or science as the king, namely, statesmanship. In other words, whether one is talking about a household or a state, the science of governing is the same. But this then means that the master of a household might share in the "kingly science" even though he is not at all a king—that is, what everyone would *call* a king—and that someone who actually is a king or who at least holds the title of king might be completely deficient in the kingly science.[13] While this might initially seem to be an opening for an argument for democracy, for allowing anyone at all to rule, it will ultimately turn into an argument for rule by those who know and who share in the science of statesmanship. It already suggests that the true statesman—the statesman in truth or in reality—may in fact be a lay person, or perhaps even a philosopher, and so may in fact go by another name than *statesman*, while the one who is *called* a statesman may be ignorant of the science of statesmanship and so may not be a genuine statesman at all.

As always in Plato, language is conventional and the task of the philosopher or dialectician is to bring names into conformity with things—to clarify those places where several different names are used for the same thing, where the same name is used for two different things, where a name exists for something that does not exist, or where a thing exists that does not yet have a name. At the beginning of the *Sophist* (216c–217a), we recall, Socrates asked the Stranger whether *sophist*, *statesman*, and *philosopher* were three names for the same thing, three different names for three different things, or some combination thereof.

It is important to note at this point that the Stranger is here *using* more than *mentioning* this important distinction between what something is

called and what it *is*—that is, in its more readily recognizable Platonic formulation, what something *appears* to be and what it actually *is*. The fact
that the Stranger goes on to make his case for the unity of the science of
statesmanship by appealing to the very familiar analogy of the physician
(the physician is a physician due to his practicing the art or science of medicine, regardless of whether he is actually recognized as or is called a physician) demonstrates that we are here on pretty familiar Platonic territory.[14]
This argument anticipates the moment much later in the dialogue when
the true statesman is defined as the one who rules because of his *epistēmē*
or his science, regardless of whether he rules by force or consent, with written laws or not, and, most importantly, whether he is a layman or an actual
ruler.

To determine whether statesmanship should be grouped among the intellectual or the practical sciences, the Stranger uses or evokes yet another central Platonic distinction. He suggests that the king is able to do
little with his hands or his body, his *sōma*, to maintain his rule and so must
rely essentially on the "sagacity and strength [*rhōmēn*] of his soul," that is,
his *psychē* (259c).[15] By applying the lexicon of the body to the soul by means
of a trope or metaphor, by speaking of the "strength" of the human soul,
the Stranger is able to associate the soul with the intellectual sciences and
the body with the practical ones. The intellectual sciences rely upon persuasion and language, upon strength of soul, as it were, in order to obtain
the support of others, while the practical sciences depend on bodily strength
and the direct, physical control of instruments and means of force. Hence
the intellectual is opposed to the practical with regard to what is made,
but also, and especially, with regard to the instrument of the making—
namely, the soul as opposed to the body. In exercising his science, the king
or statesman uses the strength of his *psychē* rather than that of his *sōma*.
The line that is drawn between the intellectual arts and the practical ones
is thus also a line between the arts of the soul and those of the body, perhaps even, to appeal to another trope, those of the head (intellectual arts)
and those of the hands (manual arts).[16]

Hence the king, and, therefore, the statesman, the master of the household,
has an art that is more akin to the *intellectual* than the practical. We can
already see here that while the art of *diairesis* makes distinctions and marks
oppositions—between the scientific and the non-scientific, the practical
and the intellectual—it also *uses* or *assumes* other distinctions in the process, that between body and soul, for example, or what something is *called*

and what it actually *is*, distinctions that are more operational than thematic. This will become crucial, as we will see in a moment, for Plato's thinking of life.

It is at this point that the Stranger makes a crucial claim about *diairesis* itself—that is, about the quasi-art or quasi-science of *diairesis*.[17] He says that to divide the intellectual arts they must look for some "natural line of cleavage [*diaphuēn*]" within this category (259d). That is, they must try to find the place where this set or category "naturally breaks," where it has articulations, joints, just like—and this analogy will be crucial for the entire dialogue—a living being. The word the Stranger uses here to describe this "natural line of cleavage," *diaphuē*, is very rare in Plato. Apart from this instance, it is used only in the *Phaedo* as Socrates is explaining in mock Anaxagorean fashion that he is sitting in prison because, he says, his "body is composed of bones and sinews, and the bones are hard and have joints which divide them [*diaphuas echei chōris ap' allēlōn*]" (*Phaedo* 98c). The formulation is telling: a joint—a *diaphuē*—can be thought of as what joins two things, two bones, for example, but also, and that is the emphasis here, as what separates two things or keeps them apart. To the question "where does Plato draw the line?" a first answer might thus already be given: he draws it there where he believes it to have *already* been drawn in nature or by nature, by the nature of things, there where things have already been— at least virtually—separated or distinguished. While *diaphuē* can be used in Greek to refer to the natural divisions in rock or in the earth, the veins or strata in inorganic matter, Plato seems in both the *Phaedo* and *Statesman* to have in mind the divisions or, better, the bifurcations in *organic* things, the single line that divides an animal body or a nut, a walnut, for example, in two, the line that one would do well to follow in order best to dissect the body or crack the nut. If there is such a thing as Platonism in a nutshell, it would seem to be inseparable from this ability to find these lines of division or distinction in the things themselves, this ability to see where concepts naturally break. The fault-line lies not in us, as it were, but in the things themselves, even if language and categories will always be necessary to draw this line out and make what is implicit in things explicit in language.

Diairesis, a method introduced by the Stranger in the *Sophist*, thus involves dividing things down the middle, at the natural joint, along the "natural line of cleavage," as if the subject under consideration were a nut to be cracked or, better, an animal to be cut up, sacrificed or parceled out.

It is as if the conceptual field or landscape were itself a *living being*, or at least were organized *as* a living being, an animal with natural joints and divisions just waiting to be carved up, as if there were, significantly, a natural fit between the method that cuts and divides and the object that is already, at least virtually, cut up. As the Stranger will soon clarify, they must divide along these natural joints, along these already defined articulations, in such a way that each part (*meros*) will form a class (a *genos*, *eidos*, or *idea*), not just a random piece of the animal but a full body part, a complete limb. One need not carve up the category—or what we might call the conceptual organism—in exactly the same way each time, inasmuch as one's end and pedagogical purpose will, as we shall see, determine how one goes about cutting and dividing, but it is essential that each cut be made along a class line and not in a way that simply separates a part from the whole. In short, one must cut along the natural joint of a category by means of a discourse that knows how to make a proper cut.

In the *Phaedrus*, the other dialogue, in addition to *Sophist* and *Statesman*, where *diairesis* is used in an explicit fashion and under this name, we see in an even more striking fashion the kind of fit or homology between the method of *diairesis* and the object of that method. It is not only some conceptual field or category, such as the set of all sciences, that is there said to have natural joints, places of articulation or cleavage to be followed, but *speech* itself, speech itself that has or *should have* natural lines of division within it. It is Socrates, this time, not the Eleatic Stranger, who uses the term *diairesis* in the context of his critique of Lysias's speech, arguing that a speech must be organized, as far as possible, like an organism—that is, like an animal body, one that will resemble as much as humanly possible, it seems, the human body. Instead of organizing a speech in a haphazard way, as Lysias has done, "every discourse [*logon*] must be organized," says Socrates, "like a living being [*zōon*], with a body of its own, as it were, so as not to be headless or footless, but to have a middle and members, composed in fitting relation to each other [*prepont' allēlois*] and to the whole" (*Phaedrus* 264c). If *diairesis* follows the natural lines of division in the things themselves, a good or proper speech will be constructed along similar lines. It, too, will be organic or like an organic, living being. An anticipation of the critique of writing later in the dialogue, Socrates argues that Lysias's speech is neither logical nor organic, that any part of the speech could have been replaced by any other, any limb exchanged with any other. It may thus look alive, sound like a living being, like a *zōon*, but it is not.[18] It is an artificial construct, an imitation or simulacrum of these "natural divisions," these "natural lines."

It is just after this claim about speech being organized like a living, organic being that Socrates in the *Phaedrus* goes on to explain the method of division that will be part of the "true art of rhetoric." Here called *dialectics*, this true art of rhetoric involves, says Socrates, two principles, namely, *gathering together into a whole* and *dividing up into classes*. The first of these principles is "perceiving and bringing together in one idea the scattered particulars, that one may make clear by definition the particular thing which he wishes to explain." The second principle is "dividing things again by classes [*kat' eidē*], where the natural joints are [*kat' arthra, hēi pephuke*], and not trying to break any part [*meros*], after the manner of a bad carver [*mageirou*]" (*Phaedrus* 265d–e). We are beginning to cut even closer to the bone, as it were, of the theme of life. The dialectician would thus be, according to Socrates, a carver, butcher, or cook, a *mageiros*, the only true or genuine *mageiros*, it might be said, insofar as he has the ability, whether by art or science or craft, to carve up not living beings but concepts or classes that are like living beings along their natural lines of division.[19]

The method of division or *diairesis* is to be used in conjunction with the principle of combination in order to follow and accentuate the articulations or jointures between categories. Once again, these natural joints, which are called *arthra*, rather than *diaphuai*, in the *Phaedrus*, suggest that the conceptual field or landscape is organized like an animal, with joints marking the places of natural jointure or separation. It is as if—and this is the hypothesis I will develop later in this work—*being* itself were organic or were being conceived in organic terms. A good carver would be someone who does not just lop off any old part but one who uses his science or his craft to follow these natural lines of separation within the things themselves, dividing in two, down the middle, whenever possible, and when not possible into a number as close as possible to two.

Speech, then, *must* be organized like a living body because the conceptual world already *is* organized like one, indeed, as we are about to see, like a body with a right side and a left, a right side that is always at once similar and superior to the left. Socrates goes on in the *Phaedrus* to say that he has made just such a division in his own two discourses on love, finding a single principle of unreason and then dividing it down the middle into two types of madness that will lead to two types of love—two types, one true and the other not, that nonetheless go by the *same name*.

> As our two discourses just now assumed one common principle [*eidos*],
> unreason, and then, just as the body, which is one, is naturally divisible
> into two, right and left, with parts called by the same names [*homōnyma*],
> so our two discourses conceived of madness as naturally one principle
> within us, and our discourse, cutting off the left-handed part [*meros*],
> continued to divide this until it found among its parts a sort of
> left-handed love, which it very justly reviled, but the other discourse,
> leading us to the right-hand part of madness, found a love having the
> same name [*homōnymon*] as the first, but divine, which it held up to
> view and praised as the author [*aition*: the cause] of our greatest
> blessing. (*Phaedrus* 265e–266b)

There are thus two kinds of madness and two kinds of love, one human,
all-too-human, and the other divine, not unlike Pausanias's distinction in
the *Symposium*. Significantly, these two kinds of madness or love have the
same name, even though only one is true, divine love, while the other is
simply *called* love, a mere reflection, imitation, or semblance of the first.

Socrates recommends separating off the left-handed side of a division
and then following out the right-handed side until one has found the object of one's search. This is precisely the process the Stranger deploys in
the *Statesman*, cutting an idea in two, following the side that is of most
interest (and, always, of most value) and abandoning the other side—that
is, always, the left side. Socrates testifies just after these lines in the *Phaedrus*, using the term *diairesis* that we have been following:

> Now I myself, Phaedrus, am a lover [*erastēs*] of these processes of
> division [*diareseōn*] and bringing together [*synagōgōn*], as aids to speech
> and thought [*legein te kai phronein*]; and if I think any other man is able
> to see things that can naturally [*pephukoth'*] be collected into one and
> divided into many, him I follow after and "walk in his footsteps as if he
> were a god" (*Odyssey* v, 193). And whether the name I give to those who
> can do this is right or wrong, God knows, but I have called them
> hitherto dialecticians. (*Phaedrus* 266b; see also 218c)

This is high praise indeed for the dialectician as a practitioner of *diairesis*,
or of division and collection, these two processes always involving one another, it seems, even if this is not always obvious or explicit. We also see
from this passage that dialectics—the true art of rhetoric—is an art of
drawing the line, in this case, the line between itself and that with which
it is often confused, namely, rhetoric as it is commonly understood and
practiced, a left-handed or deceptive form of rhetoric. While *diairesis* is

clearly not the whole of dialectic and is alone inadequate, it is hard to conclude that Plato considers it, as Michel Foucault maintains, completely
"sterile." It is true that the Stranger will himself find fault with the results
of the initial *diairesis*, necessitating the turn to myth; it is also true that he
will develop later in the dialogue—in a passage we will turn to briefly in
just a moment—a non-binary form of *diairesis*, as he divides up all the
things in the state into seven categories rather than two. But this initial
form of *diairesis* is never simply or wholly disparaged or dismissed in this
or any other dialogue and, as I have argued, it is operative in many other
Platonic dialogues, even if not under this name.

Now the reference in the *Statesman* we saw just a moment ago comparing the line of division to a natural line of cleavage, to a *diaphuē*, is echoed
much later in the dialogue when the Stranger suggests that in order to distinguish causes from contingent causes they will have to divide up all the
things within the state into a number of classes as close as possible to two
and that, to do this, they will need to "divide them like an animal that is
sacrificed, by joints [*kata melē*]" (287c),[20] the number of joints or cuts being determined, it seems, by the object itself. The result will be seven categories or classes, defined, it appears, in terms of their natural or technical
origin along with their use or role in the state. Again, it is as if the conceptual body (as if *being itself*) were organized like a living or organic body
just waiting to be carved up—that is, sacrificed and cut in pieces.[21] The
conceptual field is perhaps less like a walnut, in the end, with a *diaphuē* separating it right down the middle, and more like a body, a human body even,
which can be divided not just down the middle but into a certain number
of body parts or limbs. When the Stranger speaks of animal bodies being
sacrificed or cut up, when Socrates in the *Phaedrus* and elsewhere speaks
of speech or of *logos* being organized like a living body, it is in both cases
an anthropoid body that they seem to have in view.

In these opening pages of the *Statesman*, the goal of *diairesis* would seem
to be, as we have seen, to cut up the conceptual body or animal that would
be the set of all sciences, all *epistēmai*, into a series of parts that would also
be classes until they are then able to separate off the statesman's art or science from all the others. The Stranger thus goes on to divide the intellectual arts into those that, while not being exactly hands-on, while not
requiring one to use one's body as much as one's soul, nonetheless involve
the practitioners in the execution of their projects. Architecture is given
as an example of such an art, and it is opposed to arts such as calculation,

which seem indifferent to any practice or execution. Whereas those engaged in calculation pass "judgment" on numbers and then withdraw, the architect participates in an intellectual pursuit that nonetheless requires him to command or to give orders to others.[22] The intellectual arts can thus be divided into those that, like calculation, essentially judge—"like spectators," the Stranger adds—and those that, like architecture, both judge and command. Hence the Stranger makes yet another division by relying upon the earlier distinction between body and soul, in conjunction with a new distinction between the command of others beyond or *outside* oneself and judgment by or *within* oneself, the opposition between outside and inside being yet another structuring opposition of Platonism.[23] Inside and outside, soul and body, what something *is* and what something is *called*: While none of these oppositions are explicitly argued for in the dialogue, they are all nonetheless essential to its progression.

Obviously, the statesman falls into the category of those who do not simply judge like spectators but who both judge and command (260c–e), and who command not by repeating the orders of others—a group that includes interpreters, boatswains, prophets, and so on—but those who have "the science of giving orders of one's own," a group that is "virtually nameless," says the Stranger, or at least has no commonly recognized name. Notice here, then, that the Stranger is simply following, or so it seems, the object of their "quest" (*methodos*), not worrying whether a certain class—such as the class of all those who give their own orders—has a name or not (260e). While certain classes may thus be misnamed by ordinary language, or given a name or two too many, others may be shown upon analysis or inspection to be lacking a name altogether.

Distinguishing Plato's Zōa

But then we come, finally, though as if only in passing, to the category of *life* (261b–c). The moment is much like the one from the *Sophist*, just after the reference to the gigantomachia over the nature of being, that we looked at in the introduction to this work. Within the class of those who command by issuing their own orders, there are, the Stranger now says, those who have as the object of their command "lifeless things" (*apsycha*) and those who have "living things" (*empsycha*) (261b). Unlike architecture, the art of statesmanship has to do, obviously, with living things—that is, with animals, *zōa*, a category, we come to see, that includes both humans and other animals. The art of the statesman is, therefore, says the Stranger, "more noble" (*gennaioteron*)—that is, better born, than the art of the architect insofar as its object is always living things, it being self-evident and so in need of no further justification that living things are more noble than non-living ones.[24]

Notice, then, that while the method of *diairesis* here aimed to pick out or define the nature of the statesman, not the animal or life or anything else of the sort, other categories have come into play, categories that are *used* or assumed rather than *mentioned*, argued for, or developed. It is these more *implicit* distinctions or divisions that allow one to sketch out some very simple but nonetheless significant relations with regard to life. For example, if we juxtapose the claim that statesmanship has to do with the production—the genesis—of *empsycha* (261b) and the claim made just eight lines later that it has to do with *zōa* (261c), we can conclude that these two words are more or less synonymous. So obvious is this, in fact, that the Loeb translator (Harold Fowler), actually translates both words—at once correctly and misleadingly—as "living objects."[25] Though the words are as different in Greek as their literal English translations, *living beings* and *ensouled beings*, they do seem to be for the Stranger, if not for Plato, synonyms, insofar as a *zōon* is a being endowed with or possessed by a *psyche*, a being that is *empsychon*.

If *zōon* and *empsychon* are indeed quasi-synonyms, then we can go on to say that a *zōon*, commonly translated as *animal* or *living being*, is opposed, as something that is *empsychon*, to what is *apsychon*. In the plural, then, *zōa*, living things, animals, is a quasi-synonym of *empsycha*, ensouled things, and *both* of these are opposed to *apsycha*—that is, things without life or without a soul. This is all absolutely basic, to be sure, but it is not insignificant, and it gives us a glimpse not simply and not even primarily into the Stranger's argument but into Plato's conceptual architecture or his linguistic matrix. Insofar as these distinctions are, to use this problematic but helpful distinction once again, more *used* than *mentioned*, evoked almost in passing, we have a better chance of catching a glimpse of something that is less vulnerable to counterclaims about the way in which the dialogical, dramatic, or ironic framing of Plato's dialogues compromise the possibility of identifying any kind of a Platonic *signature*. It is this kind of analysis that also allows one to read *across* dialogues in order to ask the question of whether, for example, there are passages in *any* of Plato's dialogues that would disturb not only this opposition but this *hierarchy* between the living, the ensouled, on the one hand, and the non-living and non-ensouled, on the other, with the former being always "more noble," better born or of a better lineage, than the latter. Again, these are rather banal, non-contentious claims when considered individually, but when taken together they begin to provide a certain profile for the notion of life in Plato, its vital statistics, as it were, the values with which it is associated and those to which it is opposed (the soul rather than the body, being rather than becoming, and so on).

But let me pursue this notion of a *zōon* just a bit more with the aid of other dialogues. The Stranger has defined by means of *diairesis* the art of statesmanship as the art of having command over living things—that is, *zōa*. But what is a living thing? What is a *zōon*? Because the theme of this work is, after all, life, and because, as we will soon see, the universe or cosmos is itself called a *zōon* in the crucial myth of the two ages in the *Statesman*, it will be essential to determine just what a *zōon* is, in anticipation and as preparation for our attempt to ask much later, in Chapter 7, that contentious question regarding the supposed difference in the ancient Greeks between *bios* and *zōē*.

The noun *zōon* has, in many Greek authors, and not just in Plato, both an extended and a more restricted sense. In a first moment, it means a creature or animal in the large or general or generic sense. The term in Plato is thus sometimes translated as *animal* (see, for example, *Gorgias* 506d) or *creature* (see *Phaedrus* 230a) or *beast* (as a synonym for *thremma*, see *Republic* 493b–c)—all perfectly good and reasonable translations. As the name itself might suggest, a *zōon* is anything that participates in *zōē* or *zēn*—that is, in life or in living. As the *Timaeus* puts it, "everything . . . which partakes of life [*metaschei tou zēn*] may justly and with perfect truth be termed a living creature [*zōon*]" (*Timaeus* 77b). What makes a *zōon* live is, from this perspective, nothing other than its participation in the essence, form, or principle of life, a point I will return to in Chapter 7.

But what causes a *zōon* to participate in life is, it seems, its being held or embodied or possessed by a soul, by a *psychē*. The *Timaeus*, once again, speaks of "that compound of soul and body which we call the 'living creature' [*zōon ho kaloumen*]" (*Timaeus* 87e). A similar definition is given at *Phaedrus* 246c, though Socrates there goes on to say that it is because of this compound of soul and body—with the emphasis being placed here more on the body than the soul—that a *zōon* comes to be called mortal, *thnēton*. The *Sophist* then brings all these terms together when it defines a

"mortal animal" (*thnēton zōon*)—a phrase that would be essentially pleo-
nastic were there not, as we will see, immortal animals (see *Timaeus* 77b)—
as an ensouled body or "a body with a soul in it" (*sōma empsychon*) (*Sophist*
246e). Hence a *zōon* is the compound of body and soul and, as a result of
this compound, it is (with just a few, enigmatic exceptions, as we will
see) both alive and mortal, alive (because of its soul) and destined to die
(because of its body). Even if all souls are immortal, as Plato seems to
argue in the *Phaedo* and elsewhere, *zōa*, the *combinations* of bodies and
souls, are mortal.

Zōa are, then, in a first moment, living beings in general, everything
from plants to animals to man.[26] But then, within this general category of
zōa as living beings, Plato will often—as in the *Statesman*—give *zōa* a more
restricted sense. *Zōa* then designate not living beings in general but animals, including humans or *anthrōpoi*, animals *as opposed*, therefore, to plants,
to *phuta*.[27] From this perspective, man is a *zōon* among other *zōa*, one kind
of animal among others, all of which are opposed to plants, which are also
considered to be alive but are not, in the more restricted sense of the term,
zōa.[28] In the *Gorgias*, Socrates gets Callicles to affirm that "man is also one
of the animals" (*ho anthrōpos hen tōn zōōn estin*) (*Gorgias* 516b), that is, that
anthrōpos is one of the *zōa*, a claim or an assumption that can be found in
any number of other dialogues.[29] It is because man is a *zōon* in just this
sense that he will need to be raised, trained, and cultivated, at least in some
ways, just like other animals, conflated with other animals to such a degree that someone like Callicles will be able to claim that right is the advantage of the stronger among both men and other animals. It is this same
conflation of humans with other *zōa* that will allow the Stranger, in the
diairesis we are following, to gather both under the category of "the breeding and nurture of living beings" (*tēn . . . tōn zōōn genesin kai trophēn*),
whether that be "the nurture of a single animal" or "the common care of
creatures in droves" (*en tais agelais thremmatōn*) (*Statesman* 261d).[30]

But if man is sometimes just a *zōon* among other *zōa*, he is also sometimes a *zōon* of a very different kind, which is why there is sometimes a
subsequent distinction and, as a result, *three* categories rather than two,
namely, plants (*phuta*), animals (*zōa*), and man (*anthrōpos*). In this case, the
human is considered an animal of such a different kind that he merits a
different or another name in addition to *zōon*. It is this difference that helps
explain the kinds of stories we hear in the *Statesman* and the *Protagoras*
where *anthrōpoi* are set apart from all other *zōa*. Though the story told in

Protagoras comes from the mouth of Protagoras and not Socrates, there is nothing in Plato that would really contradict it—including the myth of the *Statesman*. Because man was initially left unprovided for in the distribution of goods by Epimetheus, he was in danger of being completely destroyed by other animals. Initially unshod, unbedded, and unarmed, he would have been utterly destroyed were it not for the gift of the arts from Prometheus and Athena—respect and justice but perhaps first and foremost the gift of *logos*. What distinguishes *anthrōpos* from other *zōa*, or else from *zōa* in general, on this account, are these arts, along with a capacity for *logos*, such that all other animals or *zōa* can simply be called, as they are called in the *Protagoras*, *aloga*—that is, beings without *logos*, *brutes* as this term has sometimes been translated (*Protagoras* 321c).[31]

   To summarize, then, the term *zōa* can refer, first of all, to all living things, plants, animals, and humans alike, but then, in a second moment, to animals, what we call animals, including the human animal, as opposed to plants. Finally, by means of a further distinction, the term can come to mean animals *as opposed* to humans, *anthrōpoi*, who are considered to be so different from these other animals that they seem worthy of a wholly other name. As the Athenian says in the *Laws*, acknowledging, it seems, the tension between these various understandings of the animal and of man: "man [*anthrōpos*] is a tame creature [*hēmeron*]," and "while he is wont to become an animal [*zōon*] most godlike [*theiotaton*] and tame when he happens to possess a happy nature combined with right education, if his training be deficient or bad, he turns out the wildest [*agriōtaton*] of all of earth's creatures" (*Laws* 766a). Hence, all *zōa* (leaving aside for the moment the question of whether the gods, like the cosmos, can be considered *zōa*) are *empsycha*, ensouled, as well as *thnēta*, mortal, while some *zōa*, namely humans, *anthrōpoi*, are *loga*, beings with *logos*, as opposed to all other animals who are, then, *aloga*, beings without *logos*, a category that more or less corresponds to *thēria*, brutes or beasts.

Long and Short Division

Statesmanship thus has to do with commanding and, as the Stranger soon parses it, nurturing living beings or ensouled beings that—and this will be the next distinction—live not singly but in groups or herds (261d–267b). Between nurturing an individual animal and nurturing a herd, the statesman's art most clearly resembles the latter. It is at this point that the

Stranger gives the Young Socrates an opportunity to try his own hand at
dividing in two the art of caring for herds. Young Socrates thus takes up
the science of caring for herds and, knowing that men, *anthrōpoi*, will ultimately have to be distinguished from all other *zōa*, he simply lops off the
class of caring for men from caring for all other animals who are not men.

After complimenting, and surely not without irony, the Young Socrates
on making this distinction so willingly and with such courage, the Stranger
points out that the aim of *diairesis* is not to divide or make a single division
right at the outset, as Young Socrates has done, in view of the final goal.
He recalls the rule that was explained earlier, namely, that the part, the
meros, that one cuts off must itself be a species, an *eidos* (or, as he will say
later, a *genos*, the two words being used here more or less interchangeably),
and he then adds that it is better to divide a set or group down the middle
so as to find more "classes." Once again, dividing a class is not unlike carv-
ing up or even dissecting a body. Rather than beginning by cutting off a
fingertip or part of a toe and then proceeding up the arms, legs, and so
forth in small increments until the whole body has been cut up, it is better—
that is, easier, safer, but also more effective and, it seems, instructive—to
begin by dividing the body down the middle or, at the very least, by cutting it at its joints (262b). This is, it seems, the way a good butcher proceeds, whether he is carving up a body for sacrifice or consumption.

By giving the Young Socrates a chance to divide on his own, the Stranger
in effect gives himself a chance to offer Young Socrates—and thus the
reader or listener—another lesson in the art of *diairesis*, as if this dialogue
were, like so many, if not all, of Plato's dialogues, as we have repeatedly
said, as much about *method* as the object of the method. The Stranger proceeds to give the Young Socrates an example of incorrect and correct division, and the example he chooses is, not coincidentally, none other than
the example of *man*, not the individual human body, this time, but the class
of human beings, *anthrōpos* in general, an example that, in Plato, is never
innocent or without consequence. When people divide human beings in
two, says the Stranger, they often separate off the Greek race from all the
other races and give all the others the name *barbarians*, as if they all be-
longed to the same species, the same *genos*, whereas the name really means
nothing more than *non-Greeks*.[32] By giving this group a "single name, they
think it is a single species [*genos*]" (262d), though it is not. While names
are sometimes indications of classes or categories, there are some names
that amalgamate classes and others that do not pick out classes at all. The

Stranger gives the example of the name *myriad*, which is a number, and so a part of the class of number, but is itself not a class within the class of number as a whole.[33] It would be better, says the Stranger, to cut things down the middle—the class of human beings into male and female, for example, or, in the case of numbers, odd and even.[34] While *male* and *female* would be categories or classes that appear in nature and that divide the class of human beings more or less down the middle, the name *barbarians* is not the name of a class at all but a conventional distinction that sometimes passes for a class only because it is opposed in common speech to Greeks. While all classes are thus parts—the class of men, for example, is a part of the class of human beings—not all parts, such as that named by the terms *barbarians* or *myriad*, are classes.

Having given the Young Socrates another lesson in *diairesis*, the Stranger returns to the division of living beings (*zōa*) into humans (*anthrōpoi*) and all other beasts, here called *thēria* (263c). This division is made possible, it seems, by the two senses of the term *zōa* we saw a moment ago, animals in general, including humans, and animals as opposed to humans. It is this distinction that allows for the category of *zōa* to be broken up into *anthrōpoi* and *thēria*, this latter being a quasi-synonym of *aloga*. Though the Young Socrates was able to give a common name to this grouping of "other beasts"—a grouping that would correspond, if we wanted to draw the connection back to the Stranger's example, to barbarians, barbarians as opposed to Greeks—that does not mean they form a class. (In each case, interestingly, the deficiency is linguistic, the barbarian not speaking Greek, the beast not speaking at all.) The Stranger says that the Young Socrates separated off too quickly all other beasts from humans, just as some separate off barbarians from Greeks, thinking, assuming, that humans are so different from other animals and Greeks from barbarians that a natural line of division can be immediately drawn between them and their others—that is, between them and everything that is not-them, Greeks vs. non-Greeks, humans vs. non-humans. The Stranger thus goes on to attribute this haste on the Young Socrates's part to a sort of hubris or anthropocentrism. If other animals that appear to think were to give names out of the same self-pride, says the Stranger, animals like the crane might separate themselves off and put man into the same group as all the other animals that are not-cranes. It will be up to the Stranger, as we will see, to give us—though it will really take him the rest of the dialogue to do this—what I think will be a *justified* version of this same hubris or self-pride.

Instead, then, of dividing the class of herded animals into humans and all other herded animals, all other herded animals that are not-man, one should, says the Stranger, make further distinctions among herded animals. Before doing this, however, the Stranger, gives us yet another lesson in *diairesis*, pointing out that this distinction between caring for individual animals and caring for herded ones itself relies upon an unvoiced or un-thematized distinction between tame animals and wild ones, animals that are *tithasos* (a quasi-synonym of *hēmeros*) and those that are *agrios*. We saw something similar when the distinction between intellectual and practical arts relied upon the voiced but unthematized distinction between soul and body. The Stranger says to Young Socrates that by going too quickly earlier they actually lost time (264b)—the Greek version, it seems, of "haste makes waste." When they divided off single animals from herd animals, they were, in effect, says the Stranger, "hunting" (*thēreuomen*) among tame creatures rather than wild ones. Had they actually been hunting, he says, they would have followed the branch of the division leading to wild animals rather than tame ones, but because they are not actually hunting but seeking the statesman, they were implicitly following the category of tame animals. In other words, they simply *assumed* the distinction be-tween tame and wild, without making it explicit, in order to follow the former under the name of "herd animals."

As in the *Sophist*, the Stranger uses a reference to hunting *within* the dialogue in order to compare the dialogue itself to a hunt, an analogy that, no doubt, is, as always, more than an analogy. Throughout the *Sophist*, one will recall, the object of the Stranger's search, namely, the sophist, is compared to a beast being hunted—a beast who is himself at one point compared to a hunter of youth (*Sophist* 221c–223b)—and the method of division or *diairesis* is compared to a hunt, yet another analogy that suggests that the forms or categories being sought already exist in nature and so must simply be tracked down through dialectic or *diairesis*, captured and then carved up into smaller pieces. As always in Plato, there is an interplay or a mirroring between what might be called the method being used in the dialogue and the object of that dialogue.[35] As a quasi-art that makes distinctions, *diairesis* or, even more powerfully, *dialectic* has a way of insinuating itself into all kinds of discourses, taking on the characteristics of those discourses, allowing itself to be *compared* to carv-ing or to cutting, to hunting, or more simply, to the process of drawing lines.

Returning to the division of tame animals in herds, the Stranger recalls that there are reports of fish farms in Persia and goose or crane farms in Thessaly (264c), so that the class of herding animals might be divided into aquatic herding and land-herding, with this latter being then divided into the herding of flying animals and walking ones (264d). At this point, the Stranger says he perceives two possible paths to take—a short one, where a small part, a part that is nonetheless also a class, is cut off from the whole, and a longer path, where one continues to divide down the middle (265a). Because they are close to the end, says the Stranger—who can obviously *see* or *foresee* the end—they will follow both paths in succession. We thus learn here that *diairesis* allows for different paths, different approaches to defining the same subject. Though there are natural joints within and between things, these things can be divided up in different ways. Not all cuts or divisions will be equally helpful or pertinent in any particular *diairesis*. Division—*diairesis*—seems to require a certain form of *judgment*, a certain (perhaps Promethean) foresight, it might even be said, that enables one to choose between various paths in view of a final goal. Hence *diairesis* always presupposes a preliminary and unvoiced *collection* on the basis of which an appropriate *division* can then be made.

Beginning with the longer path, the Stranger says that "by nature" (*physei*)—again *by nature* or *in nature* or *according to nature*—the class of walking land-animals can be divided into the horned and the unhorned (265b), and then the unhorned, the class to which man obviously belongs, into those animals that mix breeds and those that do not, the human belonging, of course, to this latter. The Stranger next divides this class of animals that do not mix breeds, using a mathematical figure, into two-footed and four-footed animals, which leads to a joke regarding the pig as a two-footed animal. The point of the joke is to put the swineherd on the same level as the king or statesman, the former being a herdsmen of four-footed animals and the latter a herdsman of two-footed ones. This conclusion proves what was argued earlier in the *Sophist*, says the Stranger, namely, "that this method of argument pays no more heed to the noble than to the ignoble, and no less honor to the small than to the great, but always goes on its own way to the most perfect truth" (267d).[36]

Having explained the longer path, the Stranger takes the Young Socrates down the shorter one, dividing the walking class immediately into quadrupeds and bipeds, and the bipeds into feathered and non-feathered, with men falling into this latter category (267d–e). Men are thus compared to two-footed pigs in the longer path and featherless birds in the shorter one. These rather strained jokes nonetheless point to a serious problem with the *diairesis*, a problem that the myth of the two ages will be called upon to address.[37] Man has been reduced, according to both paths, to an animal among other animals, a *zōon* among other *zōa*. Missing is the understanding of man or *anthrōpos* as a special kind of *zōon*—that is, a *zōon* with reason or with logos, the only *zōon*, as we saw, endowed with logos.

Despite the jokes and, soon, the objections, the Stranger and Young Socrates do seem to have found, and by two paths, what they were seeking: the statesman is one who has an art, an *epistēmē*, that is intellectual (rather than practical), that commands (rather than just passes judgment), that gives its own orders (rather than conveying the orders of others), that orders living (rather than lifeless) beings, beings that live in a community (rather than singly), that are land animals (rather than aquatic), two-footed (rather than four-footed), and non-feathered (rather than feathered) (267a–c). Having defined the statesman, they must now, says the Stranger, availing himself of yet another analogy, take "the kingly man and place him as a sort of charioteer therein, handing over to him the reins of the state, because this is his own proper science," that is, his own *epistēmē* (266e). The statesman would thus be, according to this analogy, not only a herdsman, shepherd, or physician of the state, indeed the true and most real version of these, but a charioteer, this image of the charioteer being yet one further connection between the *Statesman* and the *Phaedrus*.[38] Once again, the quality of statesmanship as a science, an *epistēmē*, has been assumed, or rather stipulated, agreed to at the very outset of the investigation and then left unquestioned throughout. While the Stranger will soon question whether they really have isolated the statesman, it will not be this attribute that will be contested, even if, in the end, everything depends on it.[39]

To conclude this introductory chapter on the *Statesman*, let me summarize what we have seen thus far about *diairesis*, this art or quasi-art of *drawing lines*. First, when following the divisions that already seem to exist within the things themselves, one must be attentive both to these distinctions and to the ultimate goal of one's search. For the goal or object of the search will determine which distinctions are more relevant or revelatory than others. The divisions ultimately lie in the things themselves, but the path one follows is determined by the purpose of one's search, which makes certain cuts more effective or more pedagogically appropriate than others.

Second, some categories or classes may have names that are acceptable (the names *intellectual arts* and *practical arts* seemed appropriate or at least acceptable for dividing up the arts), while other classes may not have a name at all (for example, the class of those arts where one issues one's "own commands"). Sometimes we have a name, a shared, common name, but no clear or common understanding of the thing so named. That is the case here of the *statesman* and, in the *Sophist*, of the *sophist*. As the Stranger says in that earlier dialogue: "as yet you and I have nothing in common about him but the name; but as to the thing to which we give the name, we may perhaps each have a conception of it in our own minds; however, we ought always in every instance to come to agreement about the thing itself by argument rather than about the mere name without argument [*touoma . . . chōris logou*]" (*Sophist* 218b–c). The goal of dialogue would be to come to an agreement about the thing, not just the name, even if clarity regarding the thing will often come by way of clarity in one's use of names. Sometimes also, as we saw, names are mistaken for classes (for example, the *barbarian* was taken for a class even though it is really just a name given to all peoples who are non-Greek). Sometimes an art or activity can go by two names even though it is really just a single thing; the *royal art* is the same thing as *statecraft*, for example, or else, as we will see later in the dialogue, *weaving* is the same as *clothes-making*, insofar as the greatest part of weaving concerns clothes-making and so differs from it, the Stranger will say, *only in name* (see 280a). And, sometimes, we may have more than one name for the same thing—the jury still being out on whether *sophist*, *statesman*, and *philosopher* name one, two, or three different things.

Finally and most importantly for my purposes here, the claim that divisions are to be made always along "natural joints" suggests that the lines of distinction drawn in language are not arbitrary or merely conventional but are based on distinctions in and among the things themselves. Regardless of whether these distinctions are recognized, have names, or have good names, they already exist in some way "in nature" and are waiting to be identified. While the names might thus be merely conventional, the categories they pick out are not.

We will see in the myth of the two ages a world where animals are divided naturally by classes, as if the human work of *diairesis* had *already* been carried out by some demiurge or divine divider, by a God who knows exactly where (and exactly when—as we will see in Chapter 5) to draw the line. We will also see in the myth that the entire universe or cosmos is itself called a *zōon*—that is, an ensouled body—which will obviously beg the enormous question, considering everything we have seen thus far, of whether the universe is, like all other *zōa*, mortal. If it is not, we will have to ask whether there are other examples of this paradox or this oxymoron of an immortal *zōon* and what that ultimately tells us about *life*. Finally, we will want to ask whether Plato is beginning to suggest in the *Statesman* and other later dialogues that there are different levels not just of being but of *life*, a hierarchy of life, perhaps, where some things, the gods, the universe, the soul, perhaps *being itself*, will be attributed *life itself*, while other things, mortal things, *zōa* as they are typically understood, *zōa* as they are usually *called*, will be said to have what is simply *called* life, a life that is then worthy of the name *life* only to the extent that it imitates life itself or the life that is beyond life, a life that is worth living within the Age of Zeus only to the extent that it is able to imitate the life—the *real life*—of the Age of Kronos, a life that is perhaps, as I have suggested, always elsewhere, while we are in the world.

CHAPTER 2

Life and Spontaneity

Seasons. In the order of time, or rather like time itself, they speak the
movement by which the presence of the present separates from itself,
supplants itself, replaces itself by absenting itself, produces itself in
self-substitution. It is this that the metaphysics of presence as self-
proximity wishes to efface by giving a privileged position to a sort
of absolute now, the *life* of the present, the living present.

—JACQUES DERRIDA, *Of Grammatology*

If most of my reflections in the previous chapter were concentrated on the
line in its relationship to life, on the way in which the Stranger uses *diairesis*, the art of drawing lines, in order to isolate and define the statesman,
most everything in this chapter will revolve around just a single *point* in
the myth of the two ages told in the *Statesman*, that singular point in time
at the end of the Age of Kronos when, all of a sudden, the universe begins
to turn in the opposite direction and everything begins to change. The
point is indeed singular; there is no other quite like it in the *Statesman* or
in any other of Plato's dialogues. Though it is repeated at the end of each
Age of Kronos—that is, each time the Age of Kronos turns to the Age of
Zeus, for there have been, it seems, many such ages and thus many such
singular turning points—this point of transition between the two ages is
itself absolutely singular and unique. While it would seem to be paired with
another point, its twin or its double, the point at the end of the Age of Zeus
when the God or Demiurge puts his hand back on the helm of the universe so that it begins to turn again under his control, it is that first point

that will interest me most here, that singular point when the universe begins to turn apparently of its own accord, under its own control, without guidance, ushering in a new age of self-rule and self-movement for mankind and the universe as a whole, an age of independent and autonomous living, but also, as if these were essentially linked, an age of degeneration, destruction, and, ultimately, death.

If I concentrate here on just this one point, I do so because it is marked by a term that, I will argue, is anything but *one* or *singular*, a term that is already and from the very beginning double if not duplicitous. That is the term *automatos*, an adjective that is typically translated adverbially as *spontaneously*, *freely*, or *of its own accord*, an adjective that, as we will see, itself oscillates, not unlike the two ages themselves, between spontaneity and automaticity, activity and passivity, positivity and negation, presence and absence, memory and forgetting, and, in the end, life and death, or, rather, two very different conceptions of life *and* death. I will argue in conclusion that this word *automatos*, which marks the transition point between the ages of Kronos and Zeus, must itself be read in the light or in the context of what we will hear later in the dialogue regarding the superiority of science over law, and particularly over written law, of *epistēmē* over *nomos* in general, the superiority of the statesman who is present to the state and to those over whom he rules, as well as the necessity of written law in the wake of the inevitable withdrawal, absence, and death of the true statesman. In other words, the Age of Kronos will be related to the Age of Zeus in the myth we are about to read in much the same way that science (as *epistēmē* or *technē* or even as *phronēsis*) will be related to law (as *nomos*) much later, or the one true regime will be related to the six imitative ones, or, as we will see by recalling the *Phaedrus*, speech will be related to writing. Hence, the Age of Zeus will correspond, in the end, to an age of writing, an age in which the teachings of the first lawgiver are recalled only through written signs—that is, in the absence of a living lawmaker or, indeed, in the absence of a once present but now departed god. I will thus spend the better part of this chapter reviewing this myth and its terminology before turning at the end to the very suggestive claim that has been made by some scholars that Plato in this myth is rewriting or reinscribing a Heraclitean notion of *mimēsis*, one that will bring us right back to the question of life and what it means, in the case of writing or the six imitative regimes or the Age of Zeus more generally, for life to imitate "life" (or vice versa).

Problems with the Diairesis

Before turning to the myth, however, let me recall very briefly where we are in the dialogue. After a brief introduction that presents the interlocutors and links the *Statesman* back to the *Theaetetus*, which, in dramatic time, would have taken place the previous day, and the *Sophist*, which would have taken place just before on the same day, the Stranger from Elea embarked on an attempt to define the statesman. Using the method of division known as *diairesis*, a method that is introduced by the Stranger in the *Sophist* but that is also spoken of approvingly by Socrates in the *Phaedrus*, the Stranger ends up locating, hunting down, and so defining the statesman as one who exercises an art or a science, an *epistēmē*, that is intellectual, that issues commands of its own over living beings, in other words, over animals, *zōa*, that live in a community, land animals that are two-footed and non-feathered (267a–c). The statesman is, in short, the one who uses his science, his *epistēmē*, to rule over or tend to these two-footed, non-feathered animals who live a common life.

But the Stranger says he is not satisfied with this definition of the statesman's art (267c–268d). For there are, he says, practitioners of other arts besides statesmanship who would claim this very same privilege or activity, something that never happens in those arts that tend to other beasts (267d). A cowherd, for example, does everything for his cows; he is their physician, feeder, midwife, soother, and musician, and no one else besides the cowherd makes the claim that he too is a tender to cows. But when it comes to tending to men, there are several others—merchants, physicians, gymnastic trainers, and so on—who would all contest the statesman's exclusive claim to being the tender of the herd of men. Indeed these others would claim that they tend not only to the herd but to the leaders of the herd. That is because—and this is perhaps an even larger problem—the tenders of humans are themselves all humans, humans that themselves require physicians, midwives, and so on, while the human tenders of other animals can, as humans—that is, as animals of a very different kind—do everything for the animals they tend.

Hence they have not really isolated the statesman's art from that of these other human contenders and the analogy between the herders of humans and the herders of other animals has been shown to be flawed. As the Stranger puts it, their discourse, their *logos*, is not yet "right and free from error." While they have been able to outline a "sort of kingly shape [*schēma*]," they have not "perfected an accurate portrait of the statesman" or separated him off from all those contenders so as to make "him stand forth alone and uncontaminated [*katharon*]" (268b–c). That is to say, they have not yet *clarified* the statesman; they have not yet defined or isolated him—purified him in his essence. They have isolated the beastly or bodily nature of the human herd, its two-footedness, for example, but they have not shown the human to be an animal with *logos*, or an animal capable of ruling over himself—that is, an animal capable of statesmanship or any other science.[1]

The Myth of the Two Ages

It is thus necessary, says the Stranger, to find a new "starting point," a new beginning or *archē*, "and travel by a different road [*heteron hodon*]" (268d); in other words, it is necessary to change not only direction but approach, in a word, method (meta-*hodos*). Because I will argue later in this chapter that the *Statesman* is related in several important ways to a cluster of dialogues that includes *Phaedrus, Timaeus, Sophist,* and *Cratylus*, let me simply point out that this terminology of a new or another beginning is but one place of intersection between the *Statesman* and at least two of these other dialogues, namely, the *Timaeus* (see 48b) and *Sophist* (232b–c). This new path or new beginning will involve telling a "famous story" (*megalou mythou*) or a well-known myth, rather than continuing on with the *diairesis* (268d–274d).[2] This will have the added advantage, says the Stranger, of offering "some amusement" (*paidian*) to the Young Socrates, who is encouraged to pay attention as if he "were a child" (*paides*) (268d)[3]—myth, amusement, and childhood all being associated here and, just as in the famous myth of writing at the end of the *Phaedrus*, put in the service of their opposites, namely, argument, seriousness, and adulthood.

This story or myth is related to and will even help explain, says the Stranger, many stories passed down through the ages, "various portents or tokens of the gods in human affairs," for example, a "change [*metabolēs*] in the rising and setting of the sun and the other heavenly bodies" (268e). Young Socrates says he has heard such tales about the sun setting where it now rises, about the reign of Kronos and a time when humans were "earthborn [*gēgeneis*] and not begotten of one another," a reminder that these tales were well-known in Greece, or at least in Athens, and that Plato is working with, even as he is transforming, a mythic tradition that had been furnished to him by his culture. Indeed, as Vidal-Naquet reminds us, the very notion of "life in the age of Kronos," that is, a kind of golden age for mankind, was a common trope in the fourth century B.C.E. Used by "philosophical and religious sects dissatisfied by the prevailing civic order," it was, Vidal-Naquet shows, at the center of many philosophical debates.[4]

Now all these tales, and others more "remarkable" (*thaumastotera*) still, the Stranger says, are the result of a single event, a single "cause" (*aition*), which is now recalled in only fragmentary form in these various myths (269b–c). By telling of this *one* event (*pathos*), the Stranger claims, he will be able at once to explain the origin of all these disparate myths and help clarify the nature of the statesman or king—the *one* being used here, as always in Plato, to help explain the *many*.

Here is how the Stranger begins his tale—using at the outset, in this opening sentence, the term *automatos*, the word around which, as I said, this entire chapter and, in many ways, Plato's entire dialogue, will revolve:

> During a certain period [*tote men*] God himself [*autos ho theos*] goes with the universe as guide in its revolving course, but at another epoch [*tote d'*], when the cycles [*peridoi*] have at length reached the measure [*metron*] of his allotted time,[5] he lets it go, and of its own accord it turns backward in the opposite direction [*to de palin automaton eis tanantia periagetai*: "of its own accord" (BKS), "spontaneously" (SB)[6]], since it is a living creature and is endowed with intelligence by him who fashioned it in the beginning [*zōon on kai phronēsin eilēchos* (from *lanchanō*) *ek tou synarmosantos auto kat'archas*]. Now this reversal [*anapalin*] of its motion is an inevitable part of its nature [*ex anankēs emphuton*] for the following reason. . . . Absolute and perpetual immutability is a property of only the most divine things of all, and body does not belong to this class. (269c–d)

I will look later in much more detail at the strange role played by the word *automatos* in this passage. Let me simply note for the moment that the term here seems to mark something like an internal principle of life or movement, a principle of spontaneity that would be proper to living beings. When the universe is no longer being guided by the God, it begins to turn *of its own accord* in the opposite direction, since it is a living creature endowed with intelligence and, as if this were implied by that intelligence and/or that life, a capacity for self-movement.

What is called the universe, the cosmos (*kosmos*), or the heavens (*ouranos*) is thus, as in the *Timaeus*, a *zōon*, not simply a body put in motion by another, or by something outside or beyond it, by God or the Demiurge, but itself a living creature—that is, an ensouled body with what appears to be its own source of motion. As an ensouled body—as a *zōon*—the cosmos is, like all *zōa*, subject to change, destined for corruption and decay. And yet, unlike any other *zōon* we have seen, it is, as the Stranger will go on to say, immortal or capable of "receiving renewed immortality from the Demiurge" (270a).[7] The universe thus presents the paradox of being a bodily, living creature, a *zōon*, that changes and undergoes corruption and yet never dies, a living body with everlasting motion.

Perfectly aware of the problem or the paradox of an immortal *zōon*, the Stranger begins speculating about these various tales passed down through the ages in order to provide—to use the terminology of the *Timaeus*—what would seem to be the most probable or likely account (the Young Socrates will call it a "very reasonable [*mala eikotōs*]" account [270b]) of how this one *zōon* or ensouled body, which undergoes a reversal in the direction of its movement, could ever have *everlasting* movement. Because we know, first, that the cosmos, as a living body, cannot turn forever, for "to turn itself for ever [*heauto strephein aei*] is hardly possible except for the power [*dunaton*] that guides all moving things"; second, that it cannot turn in two different directions by the same God, for that would be "contrary to divine law [*ou themis*]"; and, third, that it cannot be turned by two different gods in opposite directions—a two-deity thesis that Plato or Plato's Stranger is loathe even to entertain—the only remaining alternative is that God, the one and only God, the Demiurge, turns the universe in one direction and then lets it go, allowing it then to move of its own accord, a solution that is reasonable and elegant and that allows the Stranger to attribute all degeneration or corruption (some might even say "evil") not to the Demiurge but to the nature of bodily things. As the Stranger puts it, the only remaining alternative, the only reasonable conclusion, is that

> the universe is guided at one time by an extrinsic divine cause [*theias aitias*], acquiring the power of living again and receiving renewed immortality [*athanasian*] from the Creator [*dēmiourgou*], and at another time it is left to itself and then moves by its own motion [*di' heatou auton ienai*], being left to itself at such a moment [*kata kairon*] that it moves backwards through countless ages [*pollas periodōn myriadas*], because it is immensely large and most evenly balanced and turns upon the smallest pivot. (270a)[8]

The change in direction of the universe's circular motion is thus the "event" that explains all the "wonderful" or remarkable portents mentioned earlier. Because this "reversal" is the greatest and most complete change in the universe, the Stranger says, it has all sorts of dramatic effects on the animals (*zōa*) dwelling within it, including, and perhaps especially, on human animals, who "cannot well endure many great and various changes at once" (270c). At the time of this reversal there is thus great destruction of animals and only a small portion of the human race or *genos* survives. It is, therefore, the survivors of this event who have passed down to us their "many experiences wonderful and strange" as the world spun in the opposite direction, in other words, we are given to understand, in the direction opposite its present course (270d). The greatest of all these changes in the transition from what the Stranger calls the Age of Zeus, the age in which mankind is currently living, to the Age of Kronos is that every animal—every mortal creature, every *thnēton*—immediately stops aging at the point of reversal and then actually become younger and younger in mind and body until it has become like a newborn before "wholly disappear[ing]" (271d).

Now before going any further with the myth, let me return briefly to the vocabulary of life we were looking at in the previous chapter. Because the Stranger seems to parse *zōon* by *thnēton* at 271d, we have now a third quasi-synonym to add to the two we identified earlier: animals, living beings, *zōa*, are not only ensouled beings, *empsycha*, but mortal ones, *thnēta*. Every living being, every animal, is an ensouled being, and because the souls of such beings are always found in a body, every animal is also mortal. All this, as we tried to argue in the previous chapter, seems absolutely obvious, indeed almost without need of stating. And yet, as we also claimed, it is helpful to look at such notions, however obvious they might first appear, when they are being *used* rather than explicitly *mentioned*—in other words, when they are more operational than thematic. It is particularly helpful here, I think, because Plato is giving us in this same myth a *zōon* that seems to break with this otherwise systematic identification of *zōa* with mortality, namely, a universe or cosmos that, while a *zōon*, while ensouled, *empsychon*, nonetheless lives, it would seem, *forever*. The universe is, it appears, an eternal or ever-living *zōon*, a *zōon* unlike any other, a *zōon* or an ensouled being whose ultimate destiny seems to resemble that of the soul more than that of the body insofar as it, like the soul, does not perish or die.

To explain the *genesis* or generation of animals during this mythical time, this seemingly golden Age of Kronos, the Stranger says that they were born not of one another but of the earth—a way for him to integrate into his tale the kind of stories or myths recalled in the *Menexenus* regarding autochthonous or earth-born peoples (271a) as well as Plato's own noble lie from the *Republic* (414c, 468e; see also *Sophist* 248c). Pushing the limits of inference just a bit further, the Stranger goes on to argue that if the old grow younger in the Age of Kronos, then it is only natural that the dead should "come to life again" (*anabiōskomenous*) (271b)—this same word being used at *Phaedo* 72d, as we will see in Chapter 7, to explain the way in which souls, after having been separated from the body, enter into and come to animate or to live again in another body. The "earth-born" race thus means not only autochthonous but, in some sense, re-born, resurrected; the dead are re-born from out of the earth, brought back to life, reanimated, not in order to continue to live as they once did but in order themselves to grow younger and younger and then disappear. By grafting the myth of Er or the myth of the afterlife in the *Phaedo* onto this account, we could further speculate that as each individual grows younger and disappears his or her soul migrates to the previous body that that soul had occupied, a body that would then itself be resurrected at an old age and then become younger and younger until it too disappears. The dead would thus come back from the dead, grow younger and vanish, according to what business majors know to be the LIFO accounting principle, the Last In being the First Out, the last body in the ground the first to be resurrected, the second to last the second to be resurrected, and so on. And this process would be limited, we might speculate further, by the number of bodies in the earth, the process ending when the very first body to have been buried during the previous Age of Zeus—excluding those who have been removed by "some other fate," perhaps to the Isle of the Blessed—has been resurrected during the Age of Kronos.

The "memory" of this earth-born race was thus preserved by our earliest ancestors, says the Stranger, inasmuch as they were born close in time to the last reign of Kronos. These ancestors were, so to speak, "the heralds of these stories which are nowadays unduly disbelieved by many people" (271b). We have here, then, a combination of linear and cyclical time; both the ages of Zeus and Kronos move "forward," as it were, in a quasi-linear fashion, but both have a certain limit, at which point there is a return—in quasi-cyclical fashion—to the previous period. Since it is suggested that this reversal has happened many times, it is as if these many cyclical reversals were themselves contained in an even longer—indeed an indefinitely long—linear time.

It is here, then, in his description of life during the Age of Kronos, that the Stranger will use, no fewer than four more times, the pivotal term we just saw him use to describe the movement of the universe during the Age of Zeus, this strange adjective, *automatos*. Typically translated adverbially as *spontaneously, of itself, of its own accord*, *automatos* comes from *autos*, of course, meaning *self*, and, according to Liddell and Scott, the root verb *maō*, meaning to *strive after* or *long for*, to *desire eagerly, press forward, wish or claim to be*. The word has a long and fabled history, from Homer and Hesiod to Aristotle, who, in the *Physics*, relates it to the adverb *matēn*, meaning *in vain, idly, foolishly, without purpose, senselessly, at random*. It is therefore identified by Aristotle with everything that happens outside some purpose or cause, everything that happens accidentally, *tuchē* or fortune being the name given to accidental things that happen, for either good or ill, exclusively to man.[9]

In Plato, however, or at least in the *Statesman*, this word *automatos* seems to play a particularly strange and unique role.[10] It seems to describe or to characterize, as we saw in the very beginning of the Stranger's tale, a principle of life and autonomy, a kind of spontaneous or self-generated movement, a principle, perhaps, of life or life itself, of life and of movement. The subsequent uses of the term in the same myth initially appear to substantiate this meaning. For example, the Stranger goes on to say that, in the Age of Kronos, the age in which men grow younger rather than older, "all the fruits of the earth sprang up *automata* for men," that is, as Fowler translates it, "*of their own accord* for men" (*peri panta automata gignesthai tois anthrōpois*) (271d; "of itself" [BKS], "spontaneously" [SB]). Just as men during the Age of Kronos rise up out of the ground unbegotten, with no seed having to be sown in the earth or in other human beings, so the "fruits of the earth" emerge from out of the earth spontaneously, without agriculture or human toil. This is just one of those features of the Age of Kronos that distinguish it from what we might initially think to be a mere reversal or unwinding of the Age of Zeus, a sort of film in reverse.[11] In the Age of Kronos, everything having to do with creation or genesis—as well as governance, and this is what will be crucial to the Stranger's tale—is transformed, not simply reversed. Just as we do not see men in the Age of Kronos growing younger and younger and returning to their mothers' wombs, so we do not see humans in the Age of Kronos pulling seeds out of the ground, un-plowing their fields, or walking backwards. We see instead fruit offering itself of its own accord, freely, to mankind, as if spontaneously produced by some principle of life.

During this period, the Stranger continues, it was God himself who ruled over and supervised the whole revolution, with, he adds—and this addition is crucial—the various "parts of the universe divided by regions among gods who ruled them" (271d), all the animals of the cosmos being then "distributed by species [*kata genē*] and flocks [*agelas*] among inferior deities as divine shepherds [*nomēs theioi dieilēphesan daimones*]" (271d).[12] If what is being depicted here is a quasi state of nature, it differs in several significant respects from the one depicted by, say, Rousseau. Whereas both characterize this as a time before or without agriculture, Plato does not envision animals wandering about, more or less loose and separate, as Rousseau does, but already grouped together, organized, it seems, by species, separated into seemingly natural divisions. One might be tempted to compare these "natural divisions" among species and flocks to the famous image of the aviary in the *Theaetetus*, where knowledge is divided into flocks or species of birds that represent certain types or "varieties of knowledge" (*Theaetetus* 197d). But we might also compare this division by species to something we talked about at length in the previous chapter, namely, to those natural divisions or joints that were being sought and then clarified by means of *diairesis*, distinctions or divisions that, we might think, are *already* inscribed in the differentiation and distribution of animals in the Age of Kronos, differences between hoofed and non-hoofed animals, for example, differences, divisions, distinctions—a form of archē-writing, if you will, and this will become significant for what follows—that will need to be *rediscovered* during the Age of Zeus. It is, in short, as if the universe in the Age of Kronos had already undergone *diairesis* and the beings within it separated into classes or categories, so that the task of dialectics or of philosophy is to bring about an *anamnesis* of this *diairesis*, a recollection through signs—signs that are always conventional, always in time—of these immediate or natural divisions and differences. It is as if humans in the Age of Zeus—an age of difference and language—were being called on to recollect, through human intelligence and with the help of human divisions, a world of purely *natural differences*, a world of naturally *meaningful* differences that would be like the joints, the *diaphuai* or *arthra*, of a *zōon*. With this image of the earth divided up into species under the tutelage or guidance of deities, Plato would seem to be suggesting, let me say in anticipation, something like a realm of immediacy *before* language or *before* the sign, the prospect or promise of this impossible, paradoxical, or unheard of thing: a *natural* or *immediate sign* or, as we will see, a realm of life before life, a realm of *life itself*.

In the Age of Kronos, each of these lesser deities was, the Stranger now says, an "independent [*autarkēs*] guardian" of its flock. Under the guardianship or tutelage of these deities, there were no wild (*agrios*) creatures and animals did not consume one another as they do in the Age of Zeus, a wholly vegetarian life, it seems, lived among wholly tame animals. Recall that, in the initial *diairesis*, the Stranger suggested that the distinction between single and herd animals presupposed the distinction between the wild and the tame. In the Age of Kronos, where there are no wild animals and animals seem to have been grouped or herded together by species, this distinction is no longer pertinent. Each species was ruled over by a divine shepherd specific to it, it seems, and each stayed in its own sphere, never encroaching on the sphere or life of other species. There was thus neither "war" nor "strife" among creatures—that is, neither *polemos*, the kind of conflict that Socrates at *Republic* 470d–e says happens between Greeks and barbarians, nor *stasis*, the kind that happens among Greek states. With no distinction between friend and enemy, there was also, as Carl Schmitt would have remarked, no politics or statesmanship.

Now the reason for telling this tale, the Stranger reminds the Younger Socrates, was to reveal the nature of God's leadership during this Age of Kronos. As he says

> the reason for the story of the spontaneous life of mankind [*automatou peri biou*] is as follows: God himself was their shepherd [*theos enemen autous autos*], watching over them [*epistatōn*], just as man, being an animal of different and more divine nature [*theioteron*] than the rest, now tends to the lower species of animals [*alla genē phaulotera hautōn nomeuousi*]. (271e)

This analogy, as well as the hierarchies it establishes, is telling. Because analogy is the very cornerstone of the Platonic edifice, the divided line of the *Republic* being just one of the most prominent examples of this, it is worth taking a moment to spell out the implications of such an analogy. Insofar as man is the most divine or god-like of animals, he is to animals in the Age of Zeus what the gods are to man (and the other animals) in the Age of Kronos. It will thus be man's prerogative in the Age of Zeus to rule over animals in the same way that the deities ruled over him (and the other animals) in the Age of Kronos. The same analogy is found in the *Laws*, there, too, in the context of a description of the Age of Kronos. Just after a reference to "spontaneous growth" in the golden Age of Kronos, the Athenian explains:

> Kronos was aware of the fact that no human being . . . is capable of having irresponsible control of all human affairs without becoming filled with pride and injustice; so, pondering this fact, he then appointed as kings and rulers for our cities, not men, but beings of a race that was nobler and more divine, namely, daemons. He acted just as we now do in the case of sheep and herds of tame animals: we do not set oxen as ruler over oxen, or goats over goats, but we, who are of a nobler race, ourselves rule over them. In like manner the God, in his love for humanity, set over us at that time the nobler race of daemons. (*Laws* 713d)

As we saw in the previous chapter, mankind is a *zōon* like other *zōa*, but being of a more noble nature he rules over animals in the Age of Zeus just as the gods ruled over him in the Age of Kronos. It's a crucial analogy that gets extended even further later in the *Laws*: "With the return of daylight," says the Athenian, "the children should go to their teachers; for just as no sheep or other grazing beast ought to exist without a herdsman, so children cannot live without a tutor, nor slaves without a master. And, of all wild creatures [*thērion*], the child is most intractable" (*Laws* 808d). The analogical structure here becomes even more complex, more self-referring and self-reinforcing: The gods are to men in the Age of Kronos what men are to animals, or teachers to students, or masters to servants, in the Age of Zeus. All these terms will return later in the dialogue—all of them, as we will see, in relation to a discourse about life.

Tended to by God or by the gods, humans during the Age of Kronos led, says the Stranger, a *bios automatos*—that is, as Fowler translates it, a "spontaneous life" (*automatou peri biou*), a "spontaneous livelihood" (SB), a life that "came about of itself" (BKS), in other words, an easy, effortless life that seems to come as free and easy as the fruit that emerge *automata* from out of the earth during this time. Again it is tempting to think that we have here something like an origin or source of life itself, a sort of *élan vital*, a *spontaneous* life that emerges from out of itself, a life that is freely given by the earth or that gives itself without effort, that accounts for its own movement or emergence.

In the Age of Kronos, all animals—including humans—lived under God's care, or under the gods' care; humans came to life from out of the earth with no memory of their previous lives and with all their family ties effaced. Without states or families, wives or children, they came to life in a way that is reminiscent of the myth of the metals in the *Republic.* As in the myth told in the *Protagoras,* they lived out in the open without any of the supplements that would later secure their survival—without clothes, bedding, dwelling, weapons, or agriculture. As the Stranger goes on to specify, the earth furnished fruit "without help from agriculture" (*oukh hypo geōrgias*)—that is, it did so "of its own accord" (*all' automatēs anadidousēs tēs gēs*) (272a), as Fowler translates *automatēs* here, meaning, it seems, "of itself" (BKS), "spontaneously" (SB), or again, "freely," as Fowler translates the same word just a couple of pages later (274c), a fifth and final occurrence of the word in as many pages.

One explanation for Plato's repeated use of the word *automatos* in his telling of this myth of the two ages is that this word plays a central role in Hesiod's *Works and Days* and, presumably, in other accounts of a golden age based on Hesiod. In *Works and Days* we thus hear of a "golden race [*genos*] of men who lived in the time of Kronos" and who

> lived like gods without sorrow of heart, remote and free from toil and
> grief: miserable age rested not on them; but with legs and arms never
> failing they made merry with feasting beyond the reach of all evils.
> When they died, it was as though they were overcome with sleep, and
> they had all good things; for the fruitful earth unforced [*automatē*]
> bare them fruit abundantly and without stint. They dwelt in ease and
> peace upon their lands with many good things, rich in flocks and loved
> by the blessed gods. (112–120)[13]

Just as in Hesiod, then, to whom Plato is no doubt making reference and rewriting for his own ends, the earth furnished fruits during the Age of Kronos of itself, of its own accord, freely and spontaneously. The Athenian of the *Laws,* no doubt referring to the same tradition being referred to in the *Statesman,* recalls that "tradition tells us how blissful was the life of men in that age, furnished with everything in abundance, and of spontaneous growth [*automata*]" (*Laws* 713c).

The question that now arises in the dialogue is whether this life—this *bios,* this spontaneous, easy, freely given life—was for humans in the Age of Kronos "more blessed" than life in the Age of Zeus (272b). The choice would be easy, says the Stranger, if they could be assured about the nature of the desires of the people during the previous age. That is, if these people, having "all this leisure [*scholēs*] and the ability to converse not only with human beings but also with beasts [*thēriois dia logōn*]," made use of these opportunities to practice philosophy—a philosophy, let me note in passing, that would have to be motivated by something other than eros or death, since neither of these existed in the Age of Kronos—then they would have been much happier and more blessed.[14] But if these people just "ate and drank till they were full and gossiped with each other and the animals [*ta thēria*]," then life is surely better in the Age of Zeus (272c). That is, to cite *Timaeus* 73a, if men spent their time simply grazing like cattle, "with insatiate appetite," "rendered devoid of philosophy and of culture," "disobedient to the most divine part [they] possess," they would be condemned to a life that is far inferior to the one that is possible in the Age of Zeus.

Let me recall here what we saw in the last chapter regarding the term *thēria*, used here to designate beasts or animals other than man. The term is more or less a synonym for *aloga*—that is, for speechless animals— as we saw in the *Protagoras*. Here in the myth of the Age of Kronos, we have the hypothesis of beasts, *thēria*, *zōa* other than man, endowed with *logos*, *thēria* that would also be *loga*, even though, everywhere else in Plato, humans, *anthrōpoi*, are the only animals with *logos*. But such a paradoxical or illogical use of *logos* is, here, perfectly appropriate and logical, inasmuch as, according to the myth, animals in the Age of Kronos were more like humans in the Age of Zeus. The turn or return to the Age of Zeus is a sort of fall or degeneration not only for mankind but for all animals. Humans, who lived like gods in the Age of Kronos, become mere humans again in the Age of Zeus, and animals, who were like humans, become animals again. The change in ages thus signals a radical redrawing of the lines between gods and humans, on the one hand, and humans and animals, on the other—one giant leap, so to speak, for all mankind and for all the animals over which he must now rule.

Returning, then, to the moment of transition between the Age of Kronos and the Age of Zeus, the moment with which he began his tale, the Stranger says that when the time for change or for the reversal had come—that is, when "all the earth born race had been used up, since every soul had fulfilled all its births by falling into the earth as seed its prescribed number of times," animating a certain number of bodies until they were all resurrected and made to disappear—the Pilot of the universe (along with all the other gods) withdrew and the universe began to turn in the opposite direction. As the Stranger puts it, "then the helmsman [*kybernētēs*] of the universe dropped the tiller [*oiakos*] and withdrew to his place of outlook [*periōpē*], and fate and innate desire [*heimarmenē te kai xumphutos epithumia*] made the earth turn backwards [*palin anestrephen*]" (272e). Thus the helmsman (the Demiurge) withdraws to a place of outlook—a place of spectatorship or overview, a *periōpē*, a place affording a wide view—and all the other gods withdraw with him. Whereas *phronēsis* and, it seemed, self-movement were earlier attributed to this *zōon* called the universe, innate desire and fate (*heimarmenē*) now seem to be what account for the universe's autonomous movement, a movement that takes place, let me now emphasize, *in the absence* of the Demiurge and the other gods—that is, after their withdrawal.[15]

It is this withdrawal, then, as Milton would have put it, that "brought death into the world, and all our woe." For "as the universe was turned back," there was a terrible collision, just as there was in the transition between the Age of Zeus and the Age of Kronos, as "beginning and end [*archēs te kai teleutēs*] rushed in opposite directions" (273a). The great earthquake that followed resulted in the deaths of many living beings, both humans and other animals. But, after a time, things calmed down and "the world went on its own accustomed course in orderly fashion [*katakosmoumenos*], exercising care [*epimeleian*] and rule [*kratos*] over itself and all within itself" (273a–b; see *Phaedrus* 246e). Instead, then, of the gods caring for and ruling over the universe and over man, the universe and man now had to care for and rule over themselves. Animals would still not care for or rule over themselves, of course, but the universe and mankind would.

What is then recounted by the Stranger is the progressive degeneration of the universe as a whole and humankind in particular in the Age of Zeus, a degeneration that corresponds in many of its aspects to the decline of a golden age of man to a silver, a bronze, and, ultimately, an iron age in Hesiod's *Works and Days*. Significantly, this decline is characterized or explained not just in terms of natural corruption and decay but as the result of the *absence* of a once-present teacher-father and the *forgetting* of his teachings. Early on in the Age of Zeus, says the Stranger, the universe went about "remembering [*apomnēmoneuōn*] and practicing the *teachings* [*didachēn*] of the *Creator* [*dēmiourgou*] and *father* [*patros*] to the extent of its power, at first more accurately [*akribesteron*] and at last more carelessly [*ambluteron*]" (273a–b).[16] Hence the Demiurge is called a *teacher* and a *father*, names that, as we will see in subsequent chapters, will be central to understanding the relationship between this myth and the passage later in the dialogue on the inferiority *and* necessity of written law. The decline or degeneration of the Age of Zeus is explained in terms of a weakening of memory and a progressive forgetting of the Age of Kronos, a forgetting of the teachings of the Demiurge during that age. The reason for this decline or this forgetting, says the Stranger, is the "material element" (*sōmatoeides*) in the universe's composition, the fact that the natural or original element was infected with "disorder [*ataxias*] before the attainment of the existing orderly universe" (273b). As at *Timaeus* 31b, where the word *sōmatoeides* also appears, the universe, even if immortal, is composed of matter, of body, and because of this bodily nature its memory gets worse and worse or weaker and weaker the longer it survives in the Age of Zeus. There, as here, the universe would have been created by the Demiurge, created not *ex nihilo*, of course, but ordered out of already existing elements. Hence the universe initially "remembers" the orderly movement given it by the Creator, but this "memory" grows dimmer and dimmer as time goes on, another way, yet again, of attributing only good to the Creator. As the Stranger goes on to explain, "from its Composer [*synthentos*] the universe has received only good things" (273b), though the elements with which the Creator had to work were corrupted or corruptible from the beginning.

It is this forgetting of the universe, then, that accounts for the "harshness" (*chalepa*) and "injustice" (*adika*) of animals in the Age of Zeus. So long as the Creator was watching over them, so "long as the world was nurturing the animals [*ta zōa*] within itself under the guidance of the Pilot [*kybernētou*]," the evil or injustice was limited if not actually put into abeyance. But when the Creator withdraws—when the universe becomes "separated" (*chōrizomenos*) from its pilot or navigator—then injustice, forgetting, disorder, and difference all reemerge and contribute to the degeneration of the universe. As the universe's memory of the order given to it by the Creator begins to fade, the "disorder" that reined before the cosmos—that is, in a time before cosmic time—begins to take over.

In the beginning of the Age of Zeus, the universe fares relatively well, as it remembers its lesson from the Age of Kronos. But as it progressively "forgets" (*lēthē*) the order of that earlier age, the teachings of the Demiurge, it begins reverting back to its previous condition—that is, to its originary, disorderly nature. Though the degeneration is staved off for a time by means of the integration of the arts into human life, it continues on until the universe is in "danger" (*kindunon*) of destruction—itself and all those living beings within it. Seeing the "dire trouble" (*aporias*) of the universe, fearing, as the Stranger expresses it in a much commented on phrase, that "it might founder in the tempest of confusion and sink in the boundless sea of diversity" (*eis ton tēs anomoiotētos apeiron onta pontoon duēi*) (274d–e)—that is, into the unlimited sea of unlikeness, as if the relations *among* all things and the self-relations *of* all things risked being absorbed into a sea of *pure and absolute difference*—the God who ordered the universe in the first place took back up his position as its helmsman. Back at the helm, he set the world back in order, "restored it and made it immortal and ageless [*athanaton . . . agērōn*]," that is, I think we are to hear, immortal and ageless *so far as that is possible* for a corporeal thing, immortal and ageless *until the next time* it is threatened with catastrophe, immortal and ageless *for a time*.

The Stranger has now more or less run through an entire cycle of the two ages, the transition from the Age of Zeus to the Age of Kronos and then the transition back again to the Age of Zeus. But he now says he wishes to go back and underscore something crucial about the transition from the Age of Kronos to that of Zeus, namely, that as animals ceased getting younger and began aging, dying, and passing beneath the earth, humans were no longer autochthonous but born of one another. It is at this point that the Stranger introduces a notion that will be central for the rest of the dialogue. He says that humans began "imitating" (*apomimoumena*) the universe in its manner of "pregnancy and birth and nurture" (*to tēs kuēseōs kai gennēseōs kai trophēs mimēma*) (273e–274a).[17] Hence the Age of Zeus is not just an age of memory and forgetting but one of *imitation*. Just as the universe has to become its own ruler or "the ruler of its own course [*autokratora*]" (274a), so all the parts within it must, by imitating it, beget, give birth to, nourish, and rule over themselves. Whereas humans in the Age of Kronos led a life that was *automatos*, they (like the universe itself) must now come to be *autokratōr*—that is, self-ruling (as well as self-generating). There is now a sort of self-reflexivity in each of the parts that did not previously exist, a self-relation or self-identity, and thus a separation or isolation of those parts from one another.[18] In the Age of Kronos, nothing is really independent or self-related; there is a sort of immediacy or transitivity where the hand of the Demiurge guides everything, either directly or through the immediate mediation of the lesser gods. But in the Age of Zeus, this natural communication and guidance between ruler and ruled, this telepathic sympathy between them, if you will, is severed, so that mankind must now care for himself by *imitating* the care the universe now exercises over itself.

Without the "care" (*epimeleias*) of the deity, having now to rely only on themselves, many humans at the beginning of the Age of Zeus were, as in the myth told in the *Protagoras*, destroyed by beasts, who were "by nature," says the Stranger, "unfriendly" (*chalepa*) or harsh and so had become "fierce" (*apagriōthentōn*) (274b). This claim about the nature of animals is instructive. While it might seem that the Age of Kronos is a natural, pre-cultural age, it is really an age where nature has been, as it were, neutralized, an age without culture where even the fiercest of animals have been "tamed," if not, according to the hypothesis that they spoke and philosophized, "cultured" and "civilized." The Age of Kronos is thus a kind of natural realm or state of nature that resembles the heights of cultivation or acculturation, the ideal of civilization, inasmuch as the animals there are tamed and shepherded by the gods, perhaps even able to converse with one another—and perhaps even philosophically—as humans do in the Age of Zeus.

At the outset of the Age of Zeus, then, humans are bereft of the various arts, just as they are in the Age of Kronos. But without the gods to care for them, what was earlier a sort of simplicity or innocence now becomes a genuine lack that threatens the very existence of the human race. "Feeble and unprotected," "without resources or skill [*amēchanoi kai atechnoi*]," says the Stranger, humans were ravaged by the beasts. Since "no necessity had hitherto compelled them [*to mēdemian autous chreian proteron anankazein*]," since the earth had previously provided food to them "freely" (*automatēs*) and without toil (274c), humans did not know how to care for themselves and were in such dire "straits" (*aporiais*), such a state of aporia, that the gods again had to intervene—just as the Demiurge had to intervene at the critical moment the universe itself was about to founder in the "sea of diversity." This explains the kind of stories we hear in the *Protagoras*, stories, says the Stranger, of the gods giving salutary gifts to mankind, fire from Prometheus, the arts from Hephaestus and Athena, seeds and plants from Demeter and Dionysus—agriculture, therefore, and metallurgy, and everything that comes with this latter, including the fashioning of bronze and iron for tools and weapons. Humans thus had to take up these arts and "direct their own lives and take care of themselves [*tēn te diagōgēn kai tēn epimeleian*]," just "like the whole universe, which we imitate [*xummimoumenoi*] and follow through all time" (274d), says the

Stranger, invoking for a second time the notion of imitation. In order to rule over themselves, humans now had to *imitate*—in a sense that, admittedly, is not altogether self-evident—the *self-relation* that the universe now entertains with itself. With no gods to watch over them, humans had to use the supplementary arts of agriculture and metallurgy and, assuming it is an art, statesmanship in order to imitate this self-relation and so rule over themselves. It is this notion of self-rule that brings the Stranger right back to the central topic of the dialogue, namely, the nature of the statesman.

Automatos *in Plato*

That, in essence, is the myth of the two ages in the *Statesman*, a truly extraordinary myth about which so much more could be said. As Charles Kahn has rightly argued, "the interpretation of the myth is inseparable from the interpretation of the dialogue as whole."[19] As I have already suggested, however, almost everything about the myth—and thus about the dialogue—hinges on our understanding of that enigmatic adjective *automatos*, which is used no fewer than five times in the story. Of those five uses, four, let me recall, are related to the Age of Kronos and only one to the Age of Zeus. The term is used to characterize the way in which fruit in the Age of Kronos sprang up, as Fowler translated *automata*, "of their own accord for men" (*peri tou panta automata gignesthai tois anthrōpois*) (271d), from out of an earth that "furnished them of its own accord" (*automatēs anadidousēs tēs gēs*) (272a), which is to say, as it is later translated, "freely" (*automatēs*) (274c), a free and spontaneous giving that justifies the Stranger calling man's life during this Age of Kronos a *bios automatos*, that is, a "spontaneous life" (*automatou peri biou*) (271e). Four times, then, the word is used to suggest what seems to be a spontaneous, self-initiating, internal cause that benefits humankind during a golden, mythical age. But then there is the fifth use of the term, the one that in fact comes first in the story, where it is said that when the Demiurge lets go of the universe it turns backward "of its own accord [*automaton*]," since it is "a living creature . . . endowed with intelligence by him who fashioned it in the beginning" (269d). In order to make sense of the dependent clause "since it is a living creature" (or, simply, "it being a living creature") "endowed with human intelligence," it is perfectly reasonable to translate *automatos* as Fowler has done—that is, as "of its own accord." Because it is a *zōon* (a living creature, an ensouled being), the universe moves *automatos*—that is, on its own, of its own accord, from out of itself. The term *automatos* would here seem to suggest the positive presence of an *interior* principle of self-motion related to the soul—an internal life force or life-source rooted in the living nature of the universe. Because the very same word is used four more times just a few lines later to describe that free and easy life of mankind during the Age of Kronos, it seems perfectly legitimate to translate *automatos* just as Fowler and others have done—that is, in this more active sense, with, so to speak, this more positive spin.

The only problem with this reading is that, as we will now see, there is *no* other passage in Plato where *automatos* is linked to the soul or to self-movement or intelligence in this way, no other passage in which the term is inflected in this fashion. Couple this with the fact that, in the Stranger's later retelling of the transitional moment from the Age of Kronos to the Age of Zeus, it is not *phronēsis*, or intelligence, that accounts for the universe's autonomous movement but *desire* and then even *fate*, and we have reason to suspect that the universe's *automatic* movement during the Age of Zeus cannot simply be attributed to its being ensouled. Indeed, if we are to follow out the metaphor of the helmsman taking his hand off the helm of the universe, it would seem that the universe in the Age of Zeus does not so much move of its own accord, motivated by a life force or principle of internal motion, but, rather, that it drifts or wanders aimlessly on its own, abandoned, like a rudderless ship, cut off from its divine helmsman. How, then, are we to translate—and first of all understand—*automatos* in these five instances? Does it mean the same thing in all five places?[20] If so, what are the implications of this for understanding Plato more generally? Is there in the dialogues a conception of living beings, ensouled beings, moving spontaneously from out of themselves, even when, indeed at the very moment when, they have been untethered or unmoored, so to speak, from their life-source or their source of intelligence—that is, from their father, teacher, and lawgiver, from their divine helmsman—the very moment when they have been left to roam about without a shepherd to guide them?

The way we translate or understand *automatos* clearly has far more than merely philological implications. Everything said thus far about life, a principle of life, the source of life and of intelligence in living bodies, seems to revolve around it. To avoid translating this term unreflectively, then— that is, automatically or spontaneously, as if it were self-evident—it is necessary to seek guidance and instruction from the other places where this word is used in Plato, especially because, as we will see, learning or instruction will be the key to understanding this term.

In the *Theaetetus*, for instance, the term *automatos* seems to suggest not a positive, automatic, or spontaneous movement, a movement undertaken independently, of one's own initiative, but, rather, a movement that is *without* guidance or direction, *without* a helping hand. It signals being left to do something on one's own, being left to one's own devices, abandoned or forsaken. When Socrates playfully suggests in that dialogue that the followers of Heraclitus speak about war to those with whom they are fighting but "utter peaceful doctrines at leisure to their pupils," Theodorus responds that these men have no pupils, for they "grow up of themselves [*automatoi*]," "each one getting his inspiration [*enthousiasas*] from any chance source [*tuchē*]" so that "each thinks the other knows nothing" (180b–c). *Automatos* is thus related, as it is in Aristotle, to chance, *tuchē*, rather than to any kind of self-motion or *phronēsis*. Like the master Heraclitus himself, Theodorus is suggesting, each of Heraclitus's followers is not a follower at all but an autodidact—with all the dangers this entails. Rather than guiding his students like a helmsman might guide or steer a ship, he leaves them like Cratylus leaves Hermogenes and the others in the *Cratylus*—in short, perplexed, given over to chance, left to fend for themselves and to find on their own whatever correctness there may be in names.[21] Having never been guided themselves, these followers of Heraclitus are incapable of guiding others. They follow their master only to the extent that they, like him, follow no one. Diogenes Laertes says of Heraclitus: "He was nobody's pupil, but he declared that he 'inquired of himself,' and learned everything from himself."[22] Whereas Theodorus has come to Athens to teach and guide young boys like Theaetetus, the Heracliteans, he complains, grow up untaught, untended, unguided. Though the criticism in the *Theaetetus* comes from Theodorus and not Socrates, it is supported by other—indeed *all* the other—uses of the word *automatos* in the Platonic corpus. What Plato seems to deplore most about the Heracliteans around him is not their doctrine, not even the form or genre in which that doctrine is expressed, though that, too, is problematic, but, inextricably related to that doctrine and that genre, the fact that they are neither oriented themselves nor able to orient others, neither freed from their own self-deception nor able to free others from theirs.

Socrates himself uses the word *automatos* in a very similar way in the *Protagoras*. Arguing against the position that statecraft or virtue can be taught like other crafts, he cites the example of men like Pericles, who, while known for their virtue, neither train their own sons in it "person-ally" "nor commit them to another's guidance," leaving them instead, says Socrates, to "go about grazing at will like sacred oxen, on the chance of their picking up excellence here or there for themselves [*ean pou automatoi perituchōsi tēi aretēi*]" (320a). Though the critique here is in the service of an argument against the sophists as self-proclaimed teachers of virtue and not against the disciples of motion, the negative valuation of *automatos* is again unmistakable. Like the followers of Heraclitus, these boys are left to roam about on their own *without* guidance or instruction. In the absence of teaching and the arts, these sons of virtuous men are left to wander about like oxen without a driver, without a goad, their education in virtue left simply to chance—chance or hazard, let me recall yet again, being one of the ways *automatos* would come to be understood in Aristotle.

While *automatos* might have thus suggested an internal, guiding princi-ple of motion, as it seems to in a couple passages of Homer, while it might have suggested a principle of natural or spontaneous growth, as it seems to in Hesiod's *Works and Days*, it appears to signal everywhere in Plato apart from the *Statesman* the lack or absence of any guiding principle, any gen-uine intelligence, whether internal *or* external.[23] Protagoras himself later argues that while the Athenians believe all citizens possess some virtue, "they do not regard [it] as natural [*physei*] or spontaneous [*oud' apo tou au-tomatou*], but as something taught and acquired after careful preparation by those who acquire it" (323c). What is *automatos*, spontaneous, is here associated with that which comes naturally, on its own, and it is opposed to that which can be taught, acquired, or prepared for, in other words, to things whose movement is oriented by some external cause or intelligence, overseen by some authority. In the *Meno*, Socrates again opposes the au-tomatic or spontaneous, that which is given freely or by chance, to what comes about through industry, effort, and skill. He speaks of someone "who became rich not by fluke [*ouk apo tou automatou*] or a gift . . . but as the product of his own skill and industry [*sophia . . . epimeleiai*]" (*Meno* 90a). Finally, in *Alcibiades I*, Alcibiades claims that Pericles, his guardian, did not "get his wisdom independently [*tou automatou*]"—as if this would be, in his eyes, a condemnation or disqualification of it—but by consorting with wise men such as Anaxagoras and Damon (*Alcibiades I* 118c).

In each of these cases beyond the *Statesman*, it is evident that Plato val-ues everything that comes through guidance, preparation, control, and intelligence, that which comes not at random or by chance but with some external cause or oversight, some care, plan, direction, or orientation.[24] Calling Heraclitus's followers *automatoi* is thus hardly innocent—and certainly not a compliment. With this one word Plato seems to suggest that Heraclitus's so-called students or followers are adrift, unguided, untaught, disoriented and, as a result, unable to guide, orient, or educate others, unable to lead or to be pedagogical, unable to turn others in the right direction. With no Socrates to guide them, no Academy to form them, they are, in short, very much like the cosmos itself in the Age of Zeus—that is, set adrift and headed for catastrophe.

This opposition between spontaneous, unguided, or haphazard movement and intelligent guidance is also central to the *Sophist*. The Stranger there gets Theaetetus to affirm that "animals, plants, seeds, roots and lifeless substances" all come about "through God's workmanship [*dēmiourgountos*]," having been "created by reason and by divine knowledge that comes from God [*ē meta logou te kai epistēmēs theias apo theou gignomenēs*]." By agreeing to this, the Stranger points out, Theaetetus is in effect rejecting the popular belief that "nature brings forth [all those things that were not before] from some self-acting cause [*apo tinos aitias automatēs*], without creative intelligence [*aneu dianoias phuousēs*]" (265c). What this then means, the Stranger will get Theaetetus to affirm, is that all things come about through some kind of intelligence and art, rather than through mere chance or some self-acting cause. What people call *natural* things are thus in fact "made by divine art [*theiai technēi*]," while things "put together by man out of those materials are made by human art." There are, therefore, "two kinds of art," the Stranger can then claim, "the one human, the other divine" (265d–e). Once again, what happens *automatos* is opposed to that which happens by art and intelligence, in this case, by divine art and intelligence.[25]

In every case besides the *Statesman*, then, that which is *automatos* is contrasted with that which is done by some kind of art, some kind of forethought, guidance, or intelligence. Already back in the *Apology*, Plato's Socrates uses the term *automatos* in a similar way. He there says that what has befallen him—his being convicted and sentenced to death—has obviously not happened without the accord of the gods. In other words, he says, what happened to him was not random or accidental but the result of some divine plan or oversight: "This which has come to me has not come by chance [*oude ta ema nun apo tou automatou genonen*]," but because, he says, "it was better for me to die now and be freed from troubles. That is the reason why the sign never interfered with me" (*Apology* 41d). Here, too, then, that which is *automatos* is understood only negatively, as that which is *without* guidance or intelligence, *without* what is best in view.

Given this constant association, in so many Platonic dialogues, of *automatos* with that which is without force or intelligence, without guidance or direction, it would be surprising if at least some of these negative aspects of the term were not haunting the *Statesman* as well. To begin to think and translate *automatos* in light of these other dialogues, one must, therefore, begin to take into account the moments of negation that surround the word on *both* sides of the point of transition between the two ages. In the Age of Kronos, mankind is described as living, as we saw, a *bios automatos* (271e), meaning, on the one hand, it seems, a life that unfolds of its own accord, naturally, spontaneously, but also, perhaps, as we have just seen, a life *without* any other intervening cause, with nothing but the tending or shepherding of the gods to guide it. Such a life can thus be described "positively" as natural or spontaneous, a life in which "all the fruits of the earth sprang up of their own accord for men" (271d), or "negatively," as a life *before* or *without* the supplement of the arts or any other human intervention. Again in the *Apology*, Socrates himself uses the term in just this way to suggest that those in the jury who wished to see him dead could have gotten what they wished *without* any intervention on their part—that is, simply by letting nature take its course. He says to those jurors who have condemned him: "Now if you had waited a little while, what you desire would have come to you of its own accord [*automatou*]; for you see how old I am, how far advanced in life and how near death" (*Apology* 38c).

If *automatos* can legitimately be translated as "spontaneously, freely, of one's own accord," it can and must *also* be understood and so translated negatively, parsed by means of a negation, so as to suggest a life in the Age of Kronos that was lived not simply *with* the gods but *without* everything that would become necessary in the Age of Zeus—that is, *without* toil, agriculture, or any of the other arts, *without* resources or skill or foresight, *without* eros or human generation and, most importantly, *without* the necessity of man having to rule over himself, in short, *without* any kind of human statesmanship.[26] The Stranger himself already begins this work of negation and translation by parsing, in effect, his own description of the earth furnishing fruit *automatē*, "of its own accord," "freely," as "without help from agriculture" (*oukh hypo geōrgias*) (272a). The word *automatos* marks the *negation* of human intervention and the arts in order to signal the divine presence of the God or the Demiurge watching over the universe during the Age of Kronos.

But if the term *automatos* already bears, in the Age of Kronos, the negative traces of an Age of Zeus that has yet to come (or that has already been), it would be surprising if it did not bear similar traces, and perhaps all the more openly or explicitly, in the Age of Zeus, an age in which human beings are said to be, at least initially, "without resources or skill" (*amēchanoi kai atechnoi*) and where it is the withdrawal of the care of the gods that leads to difference, injustice, and death. Whereas that which is *automatos* appears to be good and sufficient in the Age of Kronos—good for the universe and sufficient for human survival—since it signals what is *without* human intervention, and without the *need* for such intervention, it is dangerous and must be supplemented by the arts in the Age of Zeus, agriculture for the nourishment of man's body, metallurgy for hunting and for war, and, we might speculate, philosophy for the well-being of his soul. Once the guidance of the Demiurge is withdrawn, the universe would thus seem to move not just or not really spontaneously, on its own, of its own accord, but, precisely, *without* the intelligent guidance of God or the Demiurge. It is this negation, which marks the dominance of the bodily or material principle, that threatens the universe and the human life within it with such destruction that the intervention of the gods will again become necessary.

It is this same negation that makes it not only possible but necessary to understand the ideal in terms derived from that which is not the ideal. Because the Age of Kronos seems to function in the dialogue as a sort of ideal to be imitated by the human statesman, it must always, it seems, be *understood* within—and then *translated* into—the terms of the Age of Zeus. For the Age of Zeus is, as it were, the supplement that allows us to think, name, and describe the Age of Kronos. As Derrida writes in a related context in "Plato's Pharmacy":

> The father is always father to a speaking/living being. In other words, it is precisely *logos* that enables us to perceive and investigate something like paternity. If there were a simple metaphor in the expression "father of logos," the first word, which seemed the more *familiar*, would nevertheless receive more meaning *from* the second than it would transmit *to* it.[27]

It is in this way that the Age of Zeus, which comes after the Age of Kronos, allows us to think and to name the Age of Kronos that precedes it. The reign of the Son—and, in this context, Zeus is above all a son—is thus necessary to think the reign of the Father.

Hence the term *automatos* in the *Statesman* seems to mean not only spontaneously or of its own accord but unguided, undirected, and unoriented, set adrift, like a ship without its navigator, or a disciple without his master, or a son without his father. Whether or not another discourse, one of life and spontaneity, perhaps coming from Hesiod, perhaps from elsewhere, has been superimposed on Plato's language or allowed to echo within it, the negative valuation of *automatos* is nonetheless unmistakable. For the word first occurs, in accordance with what appears to be the prevailing meaning of the term in Plato, just as the Demiurge takes his hand off the universe, leaving it alone or on its own—bereft of his guiding presence. It can thus initially sound as if the universe turns back on its own, no longer under the guidance of the God, according to its own principle or under its own power as a living creature endowed with soul and *phronēsis*. But because, as we hear later in the myth, the universe degenerates or loses its power as it turns backward, because the power of the universe as a living creature seems to come only from its *recollection* of the teachings of the Demiurge in the previous age, this automatic movement must also be understood negatively, as a movement that is unguided, haphazard, set adrift—like a boat whose helmsman has abandoned it, or a state whose originary statesman has gone away, or a discipline whose master has vanished, or a son whose father has died.[28] For once God or the Demiurge has taken his hand off the helm, man's survival is quite literally in his own hands, in the science and industry of his hands, however much these may owe their origin to God or the gods.

To translate *automatos* systematically or spontaneously as "spontaneously," "of its own accord," or "freely," is thus, however legitimate, to translate it without regard for the negations, supplements, and language that come in the Age of Zeus. To translate it "positively" in this way is, in effect, to make *as if* that which can be understood only as the negation of a negation were originally some positive term before all negation. It is to translate it *as if* from within the Age of Kronos even though such translations are possible only in and with the terms of the Age of Zeus. It is this kind of translation, let me say in anticipation, that will allow all the terms from the Age of Zeus to be elevated and reinscribed as the real or true version of themselves in the Age of Kronos, making it possible to say, for example, that, in the Age of Kronos, God is a divine shepherd or helmsman, indeed, the one true shepherd or helmsman, the *real* or *true* version of what these are or will become in the Age of Zeus. It is what will allow one to think that this life is but a recollection or, indeed, an imitation of *life itself.*

Heraclitus and the Statesman

It is perhaps no coincidence or just a matter of chance, then, that the word *automatos* appears most frequently in the very dialogues most preoccupied with Heraclitus, namely, *Cratylus, Theaetetus, Sophist*, and, I am now suggesting, *Statesman*. For it is there used, and there is perhaps nothing fortuitous in this either, in the very pages where some scholars have seen a direct inheritance of such Heraclitean notions as the tension or relation of opposites, a theory of *mimēsis* that posits the human body as a microcosm of the cosmos itself, or a Heraclitean Great Year, which is marked on one end by a cataclysm that leads to death and destruction (as in the Age of Zeus) and on the other by a salutary intervention that brings about revival, restoration, and resurrection (as in the Age of Kronos) (see *Timaeus* 39d). Regardless of whether the *Statesman* is actually influenced by a lost treatise by Heraclitus, as some have suggested, the figure or shadow of Heraclitus is always in the background of this dialogue, it seems, just below the surface of the myth we have been reading. This becomes clear when we consider the association we have just seen of the word *automatos* with Heraclitus, though also, and even more importantly, when the *Statesman* is put into the larger context of other late dialogues, such as *Sophist, Theaetetus*, and *Cratylus*, where the figure of Heraclitus seems to motivate so many questions of teaching and pedagogy, mastery and discipleship, and the disaster that results when this teaching fails or this pedagogy is interrupted.

Let me briefly consider, then, the hypothesis of some kind of Heraclitean influence, including, perhaps, a Heraclitean notion of *mimēsis*, in this passage from the *Statesman*. To justify this hypothesis, let me begin with just a few signs. First, the relationship between dialogues: While I do not believe that anything like a definitive dating or ordering of the dialogues will ever be possible, certain family resemblances—which are no doubt often indicative of chronology—are hard to deny. For example, the myth we just read begins with the Stranger stating that he would like to begin anew, start again from the beginning, a phrase that resonates not only with *Sophist* 232b–c and *Timaeus* 48b but *Theaetetus* 179e.[29] If Heraclitus is in the background of many of Plato's dialogues, he is explicitly foregrounded in two of these dialogues, *Theaetetus* and *Sophist*, as well as the *Cratylus*. We should thus not be completely surprised to see Heraclitus exercising some kind of influence on *Statesman*, perhaps not so much because of his doctrine of flux and change but because of the genre of his discourse or the method without method of his teaching, all of which, not coincidentally, revolve around the single word *automatos*.

Aristotle is surely right when he testifies in the first book of the *Metaphysics* that Plato's philosophy was in some sense a response to the Heraclitean doctrines of change and flux taught by Cratylus. But behind Plato's reaction to these Heraclitean doctrines may be, as I have tried to suggest by looking at Plato's use of the term *automatos*, a related and just as significant disagreement having to do with the unprincipled, ungrounded, and, ultimately, unguided and unoriented nature of Heraclitean speech, a disagreement, therefore, not just over the doctrine of Heraclitus but over the genre in which it is expressed and over the way it is taught or transmitted—or not. It is in part this emphasis on the genre of Heraclitus's speech that allows Socrates in the *Theaetetus* to identify Heraclitus not only with Protagoras and Empedocles but with "the chief poets in the two kinds of poetry, Epimarchus in comedy, Homer in tragedy" (*Theaetetus* 153a), an identification with the poetic or mythological genre that is continued in the *Cratylus* (402c) and then taken a step further in the *Sophist*, the dialogue that immediately precedes the *Statesman* in Plato's dramatic chronology.

In the *Sophist*, the Stranger argues that Heraclitus not only said the same things about flow and flux as the poets but ultimately spoke like them, uttering myths or stories rather than asking and answering questions. The Stranger says, as he prepares to review previous accounts of the number and nature of realities, that all of the Pre-Socratics, Parmenides as well as Heraclitus, spoke rather *carelessly*, telling them *mythoi* as if they were children (*Sophist* 242c). All their mythological talk about principles being at war or co-existing in peace makes it impossible, he says, to take such so-called claims or accounts seriously, indeed not even seriously enough to try to refute. Such accounts were uttered, he complains, with no thought whatsoever for how they would be supported or defended, "without caring whether their arguments carry us along with them" (*Sophist* 243a). In other words, Heraclitus might have been a sage or wise man, but he could not teach to save his life—and especially not the lives of his so-called students or disciples.

The best way of dealing with these mythologizers of being is thus to question them "as if they were present in person" (*Sophist* 243d)—that is, in the *Sophist*, as if they were present like the Stranger and Theaetetus are present to one another and themselves in a dialogue that will resemble what is later called "thought," that is, the soul's silent conversation with itself (*Sophist* 246a–b). To the genre of myth-telling, the genre of adults telling stories to children, is thus opposed the genre of dialogue, a practice of questioning and answering, of submitting oneself to the elenchus, a genre that requires the *living presence* of both the questioner and the questioned. Only by making oneself present for questioning, for a dialogical encounter, only by undergoing such questioning at the hands of a teacher, will one be able to avoid what the Stranger later calls the ancients' carelessness and lack of precision with regard to names, their laziness when it comes to dividing genera or classes, *genē*, into forms or species, *eidē* (*Sophist* 267d). With neither dialogue nor the elenchus, neither *diairesis* nor the hypothetical method, to guide them, the mythologizers of being, and Heraclitus first among them, could not help but lead his followers astray.

Heraclitus was, in the end, more a myth-maker than a dialectician, more a poet than a philosopher, more an oracle than a dialectician, more a speaker of enigmas than a pedagogue. The Stranger even refers to the Heracliteans as the "Ionian Muses" (*Sophist* 242d–e)—hardly a term of praise in Plato's lexicon—and in the *Cratylus* we hear what is, I think, the fundamental, most damning critique Plato levels anywhere against Heraclitus, namely, that he was not a teacher, that he could not or did not *guide* his students, that his oracular statements undermined any genuine communication or continuity—indeed any genuine passing on of *life*—between master and disciple.[30]

Might it be, then, as Aram Frenkian has argued, that Plato in the *Statesman* is trying to rewrite through this alternation between ages, between the guided and the automatic, the controlled and the uncontrolled, not only a Heraclitean cosmology but a Heraclitean notion of *mimēsis*?[31] Frenkian has argued quite convincingly that *mimēsis* played an important role in Heraclitus's thought, even though no explicit mention of it can be found in his work as we have inherited it. Traces of such a thought can nonetheless be found, he argues, in Pseudo-Aristotle's *de Mundo*, Pseudo-Hippocrates' *de Victu*, Pseudo-Heraclitus' *Letters* V and VI, and, most importantly for us here, in Plato's *Statesman* (269a–274e). On the basis of these references, Frenkian reconstitutes a Heraclitean notion of *mimēsis* whereby human-kind, as a microcosm of the Heraclitean universe, *learns* to imitate nature or the universe through the arts.[32] Divine reason allows man to imitate nature so as to reestablish order and proportion through contraries and through the purgation of excess and deficiency. To put this into the language of *Statesman*, man in the Age of Zeus must imitate through the arts not just the universe after it has been abandoned by the Demiurge but the order that reigned during the Age of Kronos; he must learn to imitate—that is, embody, enact—the intermittent interventions of the God so as to reestab-lish order and harmony in both the world and the human body. Even Plato's notion of imitation or *mimēsis* between sensible things and ideas perhaps owes its origin, Frenkian suggests, to this Heraclitean thought of human kind, the human craftsman, imitating the divine Logos that governs all things.[33]

Whether or not we accept Frenkian's thesis, the foregoing analysis of *automatos*, which, as we have seen, intersects in several places with Plato's critique of Heraclitus, seems to point in a similar direction. What is at issue is the presence or absence of guidance, of a teacher or a father, of something to be imitated. It is, recall, precisely with these images of a teacher and father that the Stranger describes the universe in the Age of Zeus, "remembering and practicing the teachings of the Creator and father to the extent of its power, at first more accurately and at last more care-lessly" (273a–b). "Teachings," says the Stranger, even though it is hard to say that there were teachers and students in the Age of Kronos, and father, he says, even though animals and humans were, as we have just seen, born from out of the earth and not from one another or from some god in that golden Age of Kronos. Once again, terms from the Age of Zeus are being used to characterize the Age of Kronos. When the Demiurge as father and teacher withdraws, when he takes his hand off the tiller of the universe at the beginning of the Age of Zeus, the universe as a whole and mankind in particular must remember and imitate the teachings of this absent father/ teacher/navigator. The only way to stave off the catastrophe within the Age of Zeus is to remember the teachings of that earlier time. The *Statesman* myth thus explains not only the need for the arts but, already well before the subject is explicitly broached later in the dialogue, the need for law in the form of teachings or writings that will inevitably become *separated* from the father-lawmaker who first made them. Hence the withdrawal of God or the Demiurge in the myth anticipates the logic or structure laid out, or the lesson given, later in the dialogue regarding the relationship between science and law (that is, spoken but especially written law), as well as everything that is argued about speech and writing more generally in the *Phaedrus*. When the Demiurge takes his hand off the *oiax*, off the tiller, of the universe, it is time for mankind to put its hand to the *kalamos* of writing.

CHAPTER 3

The Shepherd and the Weaver:
A Foucauldian Fable

The king is not a shepherd.

—MICHEL FOUCAULT, *Security, Territory, Population*

In the myth of the two ages, the current age of political rule where men rule over other men, where they rule over themselves, is contrasted, as we have seen, with a golden age during which a deity ruled supreme over the entire universe and over mankind within it. This golden age was a time when man did not have to work for his subsistence, when a divine sovereign provided for all of man's needs, a time before agriculture when the fruits of the earth offered themselves up to man freely and without labor, a time when the sovereign watched over men like animals in a flock, a time, then, when men were ruled *like* sheep by a divine sovereign who tended his flock *like* a good shepherd, a time, it would seem, of what Michel Foucault came to call *pastoral power.*

Plato and Pastoral Power

Before continuing on, then, with our reading of the *Statesman*, I wish to offer here some reflections on Michel Foucault's analysis of this myth of the

73

Statesman in his 1977–1978 seminar at the Collège de France, *Security, Territory, Population*.[1] Foucault there argues that Plato in the *Statesman* definitively rejects the model of pastoral power, moving away purely and simply from what he had come to see as a flawed model for the statesman, namely, the statesman as shepherd, to the more adequate model of the statesman as weaver of the characters and virtues of men, an image that Aristotle will himself employ at *Politics* 1325b38–39. I wish to show in what follows that while there is indeed such a movement from the one model to the other in the dialogue, it is anything but pure or unequivocal, and for a couple of reasons. First, there is in the myth, as we have already seen, an inevitable and problematic parasitism of terms from the Age of Zeus to describe the Age of Kronos. Second and even more importantly, the Stranger emphasizes that the universe in the Age of Zeus itself *remembers* and, perhaps, *imitates* the Age of Kronos, which suggests some kind of *relationship* between what will later be characterized in the dialogue as six inferior political regimes and the one true regime that these regimes imitate.[2] Imitation and recollection are crucial terms in the myth and, indeed, in Plato's thinking as a whole. Foucault's reading of the myth, however interesting, is thus seriously limited by its failure to give any account of these themes. If the image of the statesman as weaver thus corrects or, better, *supplements* the image of the statesman as shepherd, this latter is never totally rejected or discarded.[3] This chapter thus aims not simply to refute Foucault's reading of the *Statesman* but to underscore the themes of recollection and imitation, themes that I believe to be crucial not only to the logic and structure of the *Statesman* but to Plato's understanding of life in the dialogues more generally.[4]

Plato was, of course, neither the first nor the last political thinker in the West to portray the ruler as a shepherd, but his portrayal in the *Statesman* of the sovereign as shepherd during the Age of Kronos would seem to have been a good, if not a prime, example of what Foucault in *Security, Territory, Population* called *pastoral power*. It is thus all the more curious that Foucault would claim in this seminar not only that the model of pastoral power was foreign to the Greeks but, after a relatively extended analysis of Plato's dialogues, and particularly the *Statesman*, that Plato himself provides the most strident critique or "rebuttal" of this model. I wish, therefore, to follow Foucault's attempt to exclude pastoral power from the Platonic corpus not in order to contest Foucault's general thesis regarding the advent of pastoral power in the West—even if the "example" (if not the exceptional status) of Plato surely calls for this thesis to be rethought—

but in order to offer a more complicated and, I believe, more sympathetic reading of Plato on the nature of political rule. But just so that the reader is not tempted to think that my aim here is to "save" Plato from Foucault's reductive reading of him, the opposite is really the case. As I will argue in conclusion, Foucault's reading misses not only the force and originality of Plato's dialogues but the genuine danger they represent for politics, the danger, let me say in anticipation, of a *political anamnesis* that would allow us to recollect a state in which the sovereign is indeed a shepherd—the only true shepherd—and the people are his flock, a state whose only true salvation—and, thus, true *life*—can come from a sovereign-shepherd.

Foucault's overarching thesis in *Security, Territory, Population* is that "the idea of a pastoral power, which is entirely foreign, or at any rate considerably foreign to Greek and Roman thought, was introduced into the Western world by way of the Christian Church" (*STP* 129/133). While this theme was widespread and positively valued "in the East and in the Hebrews" and subsequently became valued in the West through Christianity—particularly in places that had prepared the ground for receiving such a model, namely, small "philosophical or religious communities like the Pythagoreans"—it was not a theme, argues Foucault, in the "major political thought" of Greek antiquity (*STP* 147/150–151).[5]

It is a provocative thesis, to be sure, in light of the *Statesman*, where Plato, a central figure in Greek political thought by any standards, certainly seems, in what we just read, to portray the sovereign precisely as a shepherd and so seems to give us a good example of what Foucault calls *pastoral power*. If Foucault is to demonstrate that the theme of pastoral power or of the "pastorate" as a topic of political theory or governmentality is important for the pre-Christian East but not for the pre-Christian West, it is imperative that he minimize or even exclude its importance in Plato. That is, he must show that while the pastorate was a major theme and metaphor in Egyptian, Assyrian, and Hebrew literature, "the shepherd-flock relationship was not a good political model for the Greeks" (*STP* 136/140)—including, and perhaps especially, for Plato.

In order, then, for Foucault to maintain his thesis that the notion of pastoral power was foreign to the Greeks or that, in a refined version, it was but a marginal strain in Greek thought, a Pythagorean perversion within Greek political thought, or, in a further refinement of the thesis, a model known to Plato—perhaps through the Pythagoreans—but one that he, Plato, would ultimately reject, Foucault must explain or explain away the presence of this model in the *Statesman*. It is essential that Plato be shown to reject this model and so side with the vast majority of Greek thinkers (Pythagoreans aside) who find that "the shepherd-flock relationship was not a good political model." As Foucault goes on to ask and then answer, "Can politics really correspond to this form of the shepherd-flock relationship? This is the fundamental question, or anyway one of the fundamental dimensions of the *Statesman*. The whole text answers 'no' to this question" (*STP* 140/144). In short, as Foucault phrases it even more succinctly later: "The king is not a shepherd (*pasteur*)" (*STP* 147/150).

The lesson of the *Statesman* is, according to Foucault, that the model of the weaver, not the shepherd, is the appropriate one for the statesman, for "political action in the strict sense, the essence of the political," consists in joining together various elements within the state, just as—according to the analogy—"the weaver joins the warp and the weft." The *Statesman* would therefore be, writes Foucault, Plato's "bona fide rebuttal of the theme of the pastorate" (*STP* 146/150). But how does Foucault develop this thesis and what will it tell us about Plato more generally?

If the Greeks ultimately came to believe that the metaphor of the ruler or sovereign as shepherd "was not a good political model," that is, not a good model to designate the relationship between the gods and man or the sovereign and his subjects, Foucault has to acknowledge that the Greeks knew of this model and, on occasion, appealed to it in various ways from Homer right up to Plato.[6] Hence Foucault begins to defend his thesis by looking at all those places where the metaphor of the ruler as shepherd occurs in the Greeks but where it is does not have the significance that some commentators have granted it. It is thus well known that Homer, for example, in both the *Iliad* and the *Odyssey*, refers frequently to Agamemnon as the "shepherd of his people," the *poimēn laōn*, a designation that would seem to define him first and foremost in terms of his pastoral power (*STP* 136/140). Foucault will therefore argue that this is but a ritual appellation of the sovereign, a merely archaic or poetic use of the term, even though Plato will himself hearken back to these same Homeric designations of the ruler as the "shepherd of his people" in order to justify a model of political rule where the ruler is, just like a shepherd, of a different order or species than those he rules.

In a second series of Greek texts that invoke the metaphor of the shepherd-ruler, those of the Pythagoreans, there is often, says Foucault, a reference to the etymology of *nomos*, or law, in the *nomeus* or the shepherd who "distributes food, directs the flock, indicates the right direction, and says how the sheep must mate so as to have good offspring," and so on (*STP* 137/141). Zeus is thus sometimes called *Nomios*, Zeus the Shepherd-God. In this Pythagorean tradition, the ruler or magistrate is defined in terms not of the power he wields but the care he exercises on behalf of those to whom he tends. In this tradition, the ruler is often characterized as *philanthropos*—a designation, let it be said in passing, that Plato associated with Socrates (see *Euthyphro* 3d), and a characterization of the ruler to which we will see Socrates appeal in his refutation of Thrasymachus in Book 1 of the *Republic*. We thus find in the Pythagoreans what Foucault admits to be a coherent and consistent tradition favoring the very model of pastoral power that, according to him, was "not a good political model for the Greeks." Unable to deny the presence and importance of this model for the Pythagoreans, Foucault will argue that this tradition is a marginal one in the Greek world and so does not compromise his overall thesis regarding the significance of the pastoral theme for classical Greek thought.

Granting Foucault, for the sake of argument, his claims regarding Homer and the Pythagoreans, we are finally left with Plato. Either Plato is the great and glaring exception to this tendency among Greek political theorists to ignore or reject the pastorate as a "good political model" or, as Foucault will try to demonstrate, he does not favor such a model either, despite all appearances to the contrary. While Foucault could countenance the Pythagoreans as an exception because of their marginal status in Greek political thought, a Platonic exception would put the model of the ruler as shepherd right at the heart of Greek thought.

After noting, then, the surprising lack of such metaphors of the pastorate in Isocrates or Demosthenes, Foucault is able to conclude that "the metaphor of the shepherd is rare in what is called the classical political vocabulary of Greece." But Foucault then adds this in order to set the stage for his confrontation with Plato:

> It is rare, with the one obvious, major and crucial exception of Plato. There you have a whole series of texts in which the good magistrate, the ideal magistrate is seen as the shepherd (*berger*). To be a good shepherd (*pasteur*) is to be not only the good magistrate, but quite simply the true, ideal magistrate. (*STP* 138/142)[7]

Plato is thus an exception, but the status of this exception is still rather ambiguous. Foucault can acknowledge Plato to be an exception to the rule that the metaphor of the shepherd-ruler was not employed by mainstream political thinkers, but he must also show that Plato is *not* an exception to the general tendency to *reject* the pastorate as a good political model.

How, then, does Foucault go about explaining away the glaring exception that Plato *appears* to be to Foucault's thesis? The first mention of Plato in Foucault's seminar *Security, Territory, Population* concerns the model or metaphor of the ruler not as shepherd of his flock but as captain or navigator of his ship. Foucault does not return to this metaphor in the seminar because his sights are set from the outset on the sovereign as shepherd or as weaver. But it is perhaps not insignificant that this metaphor appears so frequently in Plato's dialogues, both early and late, oftentimes as an example of ruling over others with a certain knowledge or *technē* (see *Alcibiades I* 134e–135a). Moreover, this model of the ruler-navigator is used in the *Statesman* itself, as we have seen, when the divine ruler in the Age of Kronos is compared to the navigator of a ship who keeps his hand on the rudder or tiller of the universe until it is time for him to let it go and let the universe move of its own accord (272e, 273d–e). All this is yet supplementary proof that—and this will be my argument in conclusion—Plato never simply abandons or replaces *any* of his models but is willing to advance different *complementary* models when his dialogues demand them.

Foucault acknowledges from the outset that there are numerous references in Plato to the ruler as shepherd, multiple uses of the metaphor of the shepherd as ruler of his flock. Two hypotheses are advanced to explain these references: Either Plato had been influenced by the Pythagoreans and so had undergone an Eastern influence or this metaphor of the ruler as shepherd of the flock is a mere commonplace or traditional designation for the Greeks, just as it is in Homer. In the end, Foucault will use both hypotheses in order to explain away all the references to the ruler as shepherd in Plato.

Foucault begins by dispensing with what he believes to be the least convincing counterarguments or counterexamples to his thesis, saving for the final showdown what his readers or listeners will believe to be the strongest. Just as Homer's and the Pythagoreans' uses of the model of the ruler as shepherd were explained away in rather short order as a way to clear the field for a confrontation with Plato, so the references to the ruler as shepherd in all of Plato's dialogues other than the *Statesman* (especially *Republic*, *Laws*, and *Critias*) are grouped together and dealt with in quick succession in order to leave time for a more thorough reading of this one crucial, potentially exceptional dialogue in the corpus of this one potentially exceptional Greek thinker: "I think the *Statesman* should be examined separately . . . Let us leave it aside for a moment and take up the other texts in which Plato employs the metaphor of the shepherd-magistrate" (*STP* 138–139/142).

Leaving the *Statesman* aside for a moment, then, Foucault claims that all the other references to the ideal ruler as shepherd can be grouped into three categories. First, the metaphor is used to describe the power or reign of the gods over man in some earlier time, some golden age before man's degeneration. In these accounts, the gods are compared to shepherds who nourished, guided, and tended to men in these earlier times. As Foucault rightly says, citing Plato's *Statesman* already, humans "had no need of a political constitution" during these earlier times "because the deity was their pastor" (271e). Politics can thus begin, Foucault argues, only "when this first age, during which the world turns in the right direction, comes to an end" (*STP* 144/147–148). Foucault concludes: "The gods really were the original shepherds (*pâtres*) of humanity, its pastors. . . . This is what you find in the *Critias* (109b–c), and it is found again in the *Statesman*, and you will see what, in my view, this means" (*STP* 139/142–143). Foucault has left aside for the moment his treatment of the *Statesman*, but it is already evident how he will read the reference to the gods in that dialogue: they will be understood as mythological references to a bygone age that Plato believes we have left behind; they will thus not be understood, as several passages from the *Statesman* and other dialogues would lead us to think, as figures of an *ideal* that can never be achieved or returned to but that can and must be *recollected* or *imitated* as an ideal. Foucault's argument will be that the myth gives us a wholly negative model of statesmanship that needs to be criticized and *replaced* by another—that is, by the human statesman as weaver.

In a second series of references, the good magistrate in the time after this earlier, golden age—that is, in the time during which man must rule over himself—is also characterized as a shepherd. This alone might seem to be irrefutable evidence that Plato does indeed take the model of the ruler as shepherd as a *good* model for political rule. But Foucault argues that in these places such a magistrate-shepherd is not the founder of the city or an original lawgiver but simply a "principal magistrate" who rules over others but is himself also subordinated to others (*STP* 139/143). He is thus a "subordinate magistrate," "a functionary-shepherd," who does not represent "the very essence of the political function . . . but merely a lateral function that in the *Statesman* is called, precisely, auxiliary" (*STP* 139/143). Referring to the *Laws* (Book 10, 906b–c), Foucault concludes that this magistrate-shepherd is but an intermediary between the watchdog and the true master or legislator, a kind of middle-management shepherd and not at all an ideal or model for political rule.

The third kind of reference regarding the ruler as shepherd is to be found in the famous passage we alluded to earlier from Book 1 of the *Republic*. That is the passage in which Thrasymachus objects to Socrates's claim that the shepherd governs not for his own sake but for the sake of his flock by countering that Socrates has ignored the fact that the shepherd always tends his sheep with the end of slaughtering them and profiting from them. The real shepherd, Thrasymachus concludes, tends to his flock for *his* own advantage and not that of his sheep.

Foucault's treatment of this passage is rather curious. He begins by arguing that Thrasymachus appears to be attacking the theme of the ruler-shepherd "as if it were obvious, or a commonplace, a familiar theme at least" (*STP* 139/143), one that, I think we are to infer, Plato would have gotten from Homer or from the Pythagoreans and that he is himself already questioning if not criticizing. Hence Thrasymachus would be arguing that the "comparison with the shepherd really is not appropriate for describing the virtue necessary for the magistrate (343b–344c)" (*STP* 140/143). But this is, of course, not the end of the argument in Book 1, only the end of Thrasymachus's argument, which Socrates answers by suggesting that, in Foucault's words, Thrasymachus is "not defining the good shepherd, the true shepherd, or even the shepherd at all, but the caricature of the shepherd. . . . The true shepherd is precisely someone who devotes himself entirely to his flock and does not think of himself (345c–e)" (*STP* 140/143–144). Foucault does not underscore it, but Socrates in the *Republic* thus defends the very commonplace—or familiar theme—of the ruler as shepherd that Thrasymachus attacks. In order to explain or explain away this reference as well, then, Foucault resorts back to his earlier thesis regarding a Pythagorean influence.

> It is certain . . . well probable anyway, that this is an explicit reference, if not to that commonplace that does not seem to be so common in Greek thought, then at least to a theme familiar to Socrates, Plato, and [Platonist] circles, to the Pythagorean theme. I think it is the Pythagorean theme of the magistrate-shepherd, of politics as shepherding, that clearly surfaces here in Book One of the *Republic*. (*STP* 140/144)

It is easy to understand Foucault's hesitation at the beginning of this passage. If the theme of the ruler as shepherd is a not-so-common "commonplace" in Greek political thought, if it is a familiar theme only for the Pythagoreans, who are themselves of marginal importance to this thought, then it is difficult to explain what this theme is doing in this important passage of the *Republic*, where it is not only developed at some length but actually defended by none other than Socrates. It is at precisely this point that Foucault turns, without any further comment on this potentially damaging passage from the *Republic*, to the *Statesman*, as if this latter dialogue were able to correct the error or at least the misperception created by the earlier one, as if it were able to clear up any lingering doubts about Plato's attachment to the model or metaphor of the statesman as shepherd. Foucault says, right after referring to "the Pythagorean theme of the magistrate-shepherd . . . in Book One of *The Republic*":

> This is precisely the theme that is debated in the great text of the *Statesman*, for I think the function of this text is precisely to pose directly and head on, as it were, the problem of whether one can really describe and analyze, not this or that magistrate in the city-state, but the magistrate par excellence, or rather the very nature of the political power exercised in the city-state, on the basis of the model of the shepherd's action and power over his flock. Can politics really correspond to this form of the shepherd-flock relationship? This is the fundamental question, or anyway one of the fundamental dimensions of the *Statesman*. The whole text answers "no" to this question, and with a no that seems to me sufficiently detailed for us to see it as a full rebuttal of what Delatte called, wrongly it seems to me, a commonplace, but which should be recognized as a familiar theme in Pythagorean philosophy: The chief in the city-state must be the shepherd of the flock. (*STP* 140/144)

Thus we come to the *Statesman*, where Foucault's thesis will either stand or fall. For even if one were to grant Foucault not only his arguments about Homer and the Pythagoreans but also his interpretation of the three uses of the metaphor of the shepherd-ruler in all the other dialogues except the *Statesman*, there would still remain the *Statesman* itself, a dialogue in which the ideal ruler would certainly seem to be compared to a shepherd. Unless Foucault can explain away what looks, at least early on in the dialogue, to be a defense of the shepherd-ruler model, then this single exception in the dialogues of Plato risks making Plato a major exception in Greek political thought. Foucault's argument regarding the marginality of the model of pastoral power in the West before the advent of Christianity thus depends on Foucault demonstrating that what would seem to be a central theme of the dialogue is not one at all, or, better still, that Plato himself develops an explicit *critique* or *rebuttal* of this theme.

Relying upon some of his earlier conclusions, Foucault argues along two complementary lines. First, the references to the ruler as shepherd in the *Statesman* is evidence not of a Greek commonplace, but, again, of a Py-thagorean influence. But because this Pythagorean influence would prove deadly to his thesis if no less a thinker than Plato were to have adopted it, Foucault must go on to argue that the image of the statesman as shepherd is ultimately *opposed* by him to another, more appropriate image or meta-phor, the statesman as weaver. While Plato will indeed propose this other image or model for the statesman, the question will be whether he really does *oppose* the two and whether the one can really be said to replace or reject the other, whether the image of the statesman as shepherd is not still operating, and for essential reasons, in Plato's political ontology. The question will be, in the end, whether we should read Plato within the his-torical or epochal narrative Foucault seems to want to retain at all costs, and especially at the cost of all exceptions, or whether another reading is needed, one that reads Plato's *Statesman* not in terms of a logic of succes-sion and replacement (the shepherd by the weaver) but in terms of a logic of supplementarity that is in line with what I earlier called Plato's theory of *political anamnesis*.[8]

Foucault begins his analysis of the *Statesman* by following the Stranger's use of the art of division near the beginning of the dialogue to define the art of the statesman. At the end of a long, somewhat tedious and mechani-cal process of division, says Foucault, the statesman is defined as one who exercises the art or science of ruling over those community-dwelling ani-mals we call human beings. In short, says Foucault, "the politician is the shepherd (*berger*) of men, he is the shepherd (*pasteur*) of that flock of living beings that constitutes a population in a city-state (261e–262a)" (*STP* 141/145). But this is, as Foucault is right to point out, just the beginning rather than the end of the story or the dialogue. For "in its evident clum-siness it is fairly clear that this result registers, if not a commonplace, then at least a familiar opinion, and the problem of the dialogue will precisely be how one extricates oneself from this familiar theme" (*STP* 141–142/145). Leaving aside the ambiguity hovering here over the status of this theme in Plato's dialogue (a commonplace versus a familiar theme), Foucault is surely right to suggest that Plato will want to "extricate" himself from at least this presentation of the theme. The question then becomes *how* Plato extricates himself and how we are to understand the relationship between the successive stages of his argument.

Foucault suggests that "the movement of freeing oneself from this familiar theme of the politician as the shepherd of the flock takes place . . . in four stages" (*STP* 142/145). First, Plato begins by contesting the method of division, which is "so crude and simplistic in its first moves" and separates man off so quickly from all the other animals on the basis of a shared reference to shepherding or herding (262a–263e) (*STP* 142/145). As Foucault argues, this "theme of shepherd as the invariant is completely sterile," since we simply end up with is a series of variations based on the types of animals being tended to and there is no specificity to the statesman as shepherd of the human flock (*STP* 142/146). But what if Plato did not simply wish to *distinguish*, as Foucault assumes, but to *compare* humans and animals on this score? For it could be argued that the *diairesis* aimed to show that between the shepherd of men and the shepherd of other kinds of animals there are as many similarities as differences. While the division thus requires refinement at the level of the statesman's activity, it nonetheless sets up the homology between the shepherding of men and the shepherding of other animals, a homology that will be exploited in the myth of the two ages that soon follows.

The second moment consists in asking what exactly a shepherd is or does. What is important to note in this regard, says Foucault, is that the shepherd is always singular, always alone; there are never several different shepherds for the same herd, and this itself indicates a model of sovereignty that Plato will ultimately reject, opting instead for multiple rulers, multiple weavers. Moreover, this single and solitary shepherd takes care of a whole set of different needs and carries out a variety of functions: he feeds the herd, tends to the health of the herd, plays music for the herd, arranges unions for the herd, and so on. But when we apply this principle of singleness to the human shepherd, serious problems arise. Foucault argues:

> It is at this point that the principle of the singleness, the uniqueness of the shepherd, is immediately challenged, and we see the birth of what Plato calls the rivals of the king, the rivals of the king in shepherding. If the king is in fact defined as shepherd (*pasteur*), why not say that the farmer who feeds men, or the baker who makes bread and provides

them with food, is just as much the shepherd of humanity as the
shepherd who leads the flock of sheep to grass or gets them to drink?
The farmer and the baker are rivals of the king as shepherds of
humanity (267e–268a). (*STP* 143/147)

Whereas no one else claims to fulfill these various functions for the shepherds of animals other than men, there are among men many others besides the king-shepherd who claim to tend to the herd in one of these
capacities. Hence the farmer, the baker, the doctor, or the trainer can each
claim that he too is a "shepherd" of men insofar as he nourishes or heals or
trains the herd under his care.

Foucault is again right to claim that Plato (or the Stranger in the *States-*
man) contests the singleness of the shepherd at this point in the dialogue.
But Plato does so, it seems, only in order to clarify and perhaps even
underscore the nature of this singleness. As we will see, the ruler as a shepherd of men—as opposed to the shepherd of other animals—will indeed
need others to help him care for the state (bakers and doctors and trainers),
but this will not necessarily compromise the singularity of the human
shepherd's activity or the singularity of the human shepherd. It will instead
suggest that some other criterion besides simply tending to the flock (for
example, knowledge or science, *epistēmē* or *technē*, etc.) is needed to pick
out the particularity of the ruler-shepherd. If Plato does indeed ultimately
reject the idea that, in our political age, in the Age of Zeus, a single king
preeminent in virtue can be found to lead the city as a shepherd would lead
his flock, or as a captain would navigate his ship, this does not mean that
such models are ever completely rejected or that multiple leaders should
not always try to *imitate* the one. Plato never simply rejects the one in favor
of the many (or vice versa): not in ontology and not in politics. Foucault
summarizes where we have gotten so far in the dialogue:

> On the one hand, there is the series of all the possible divisions in
> animal species and, on the other, the typology of all the possible
> activities that may be related to the shepherd's activity in the city-
> state. Politics has disappeared. Hence the problem has to be taken up
> anew. (*STP* 144/147)

It is here that we come to the third moment, according to Foucault, in
Plato's refutation of the pastoral theme, the famous myth of the two ages,
the very myth in which Plato seems to argue that the ideal statesman not
only can but should be thought along the lines of a shepherd. As we have

already suggested, Foucault's entire argument will depend on his ability to demonstrate that this myth gives us not a positive model for the statesman, a model that may need to be refined or combined with others but that remains nonetheless intact, but a negative model that needs to be replaced by another. Let me allow Foucault to recount the myth of the two ages so that we can see precisely what he leaves out of his account:

> This is the idea that the world turns on itself, first in the right direction, or anyway in the direction of happiness, the natural direction, and then, when it has run its course, this is followed by a movement in the opposite direction, which is the movement of difficult times (268e–270d). Humanity lives in happiness and felicity so long as the world turns on its axis in the first direction. This is the Age of Kronos. This is an age, Plato says, "that does not belong to the present constitution of the world, but to its earlier constitution" (271c–d). What happens at this point? There is a whole series of animal species and each one appears as a flock. At the head of this flock there is a shepherd. This shepherd (*berger*) is the divine pastor (*génie pasteur*) who rules over each of the animal species. Among these animal species there is a particular flock, the human flock. This human flock also has its divine pastor. What is this pastor? It is, Plato say, "the deity himself" (271e). The deity himself is the pastor of the human flock in this period of humanity that does not belong to the present constitution of the world. What does this pastor do? In truth, his task is infinite, exhaustive, and, at the same time, easy. It is easy inasmuch as the whole of nature provides man with everything he needs: food is provided by the trees; the climate is so mild that man does not have to build houses, he can sleep beneath the stars; and he is no sooner dead than he returns to life. And it is this happy flock, with abundant food and endlessly living anew, this flock without dangers or difficulties, over which the deity rules. The deity is their pastor, Plato's text says, "because the deity was their pastor, they had no need of a political constitution" (271e). Politics begins, therefore, precisely when this first age, during which the world turns in the right direction, comes to an end. (*STP* 144/147–148)

That's pretty much everything Foucault notes about the Age of Kronos, a description that is relatively accurate, as far as it goes, but somewhat incomplete. We noted in the previous chapter, for example, the way in which living beings in the Age of Kronos were *already* divided up by species, as if they had already undergone the kind of divisions the Stranger attempts at the beginning of the dialogue—that is, as if they had already been divided into *eidē*, with each species of animal being tended to by a different divinity and these various divinities being overseen by the Demiurge who acts as the shepherd for mankind and for the universe as a whole. Plato may indeed be suggesting that the gods ruled in this way only in a mythical past that will not return and perhaps never was in the first place. But what is one to make of the fact that Plato goes on to *compare* this situation in the Age of Kronos to that in the Age of Zeus: "God himself was their shepherd, watching over them, just as man, being an animal of different and more divine nature than the rest, now tends the lower species of animals" (271e). This analogy is crucial, but Foucault says nothing about it. For Foucault, one age simply follows upon another, and with the change in age comes a change in the model of political rule. But Plato's analogy between God as shepherd over humans in the Age of Kronos and humans as shepherds over other animals in the Age of Zeus suggests not only a linear narrative where one model replaces another but a supplementary and *analogical* structure where terms are not abandoned or rejected but integrated into a more complex relationship. As we have seen, man is, in the Age of Kronos, a *zōon* like other *zōa*, and so he is ruled like one. But because man is also the most divine of all animals, he rules over these other animals in the Age of Zeus just like the divinity ruled over him in the Age of Kronos. Plato's analogy would seem to suggest that even though the Age of Zeus comes after the Age of Kronos, it is not only legitimate but necessary to think the earlier age in the terms of the later one. It is the image of a divine shepherd ruling over man in the Age of Kronos that provides the image of a human shepherd ruling over animals in the Age of Zeus, but it is also, as we saw in the previous chapter, the *name* shepherd and the *analogy* between the shepherd and the ruler in the Age of Zeus that allows the Stranger to characterize the divine ruler as a shepherd in the Age of Kronos.

Plato's myth must thus be read on more than one level. When the myth is read in conjunction with other parts of the dialogue and with other dialogues, it becomes clear that this age of the rule of the divinities refers not only to some mythological, golden age that is now past but to an aspect of humankind in the present. As the most noble and divine of animals, man resides somewhere between animal and divinity and so must be thought in terms of *both* ages. While Plato, later in the dialogue, will use the image of the weaver to correct the mistaken impression given by the shepherd that the ruler is of a different species than the ruled, it remains the case that the ruler, in the best of the inferior states, is god-like or more god-like than those over whom he rules. As Plato will suggest in the *Laws* and elsewhere, there is a certain part of man, the most divine part, that is destined to rule over the other, less divine, more animal parts, just as, it seems, the most divine men in the polis should rule over the less divine (*Laws* 714a–b).

That this narrative of the two ages is not to be thought as a mere progression from one age to the next, as a linear narrative that runs from the Age of Kronos to the Age of Zeus, from the age of the shepherd to the age of the weaver, is underscored already in the *Statesman* by the fact that the transition from the Age of Kronos to the Age of Zeus is not a one-time event but a cycle wherein the Age of Zeus eventually returns to the Age of Kronos (something Foucault fails to account for)—as if Plato wanted us to think both the differences *and* the similarities between these two ages.[9]

Foucault's conclusion regarding politics in this first part of the myth is nonetheless correct: insofar as the divine shepherd does everything for man, a "task [that] is infinite, exhaustive, and, at the same time, easy" (*STP* 144/148), man had no need for a political constitution. Politics begins only after this first, idyllic time, only after the world begins to turn backwards and the divinities, who are no longer "everywhere and immediately present" as they were in the Age of Kronos, help man to help himself through those supplements called the arts (*STP* 145/148). Here is how Foucault tells the second part of the story:

> Politics begins when the world turns in the opposite direction. When the world turns in the opposite direction, in fact, the deity withdraws, and difficult times begin. For sure, the gods do not completely abandon men, but they only help them in an indirect way, by giving them fire, the [arts], and so forth (274c–d). They are no longer really the shepherds who were everywhere and immediately present in the first phase of humanity. The gods have withdrawn and men are obliged to direct each other, that is to say, they need politics and politicians. However, and here again Plato's text is very clear, these men who are now in charge of other men are not above the flock in the way in which the gods are above humanity. They are themselves a part of humanity and therefore cannot be seen as shepherds (275b–c). (*STP* 144–145/148)

Plato's text is indeed "clear," as Foucault argues, but only if we take it at face value and assume that the lines of division between divinity, man, and animal can be clearly and distinctly drawn—that is, only if we assume that Plato's distinctions are not made *analogically* and that when Plato speaks of a certain divinity or a certain animality in man he is speaking only *metaphorically*. For what is, after all, Plato's tripartite theory of the soul in the *Republic* and elsewhere if not precisely a thinking of man as somehow an intermediary between the animal or beast and the divinity or true sovereign? Foucault tends to ignore all the analogies in Plato's dialogues that establish a relationship between the divine and the human, the Age of Kronos and the Age of Zeus, analogies that are, on my reading, the very backbone or lifeblood of Plato's thinking. Moreover, if those men in the Age of Zeus who are "in charge of other men are not," as Foucault argues, "above the flock in the way in which the gods are above humanity," if "they are themselves a part of humanity and therefore cannot be seen as shepherds" (*STP* 145/148), then what are we to make of the fact that these men are nonetheless expected to *imitate* the divine shepherd of the Age of Kronos? Foucault does not address this question of imitation at any point in the seminar, even though it is precisely that which, on Plato's account, links the Age of Zeus to the Age of Kronos. I will return to this theme in a moment, but not before we look at how Foucault develops the fourth and final stage of his argument regarding the *Statesman*.

The fourth stage in Plato's critique of the model of the statesman-shepherd is, according to Foucault, the moment when politics and politicians become necessary. It is at this point that the statesman can no longer be considered a shepherd but, instead, a weaver (279a–283b). This "model of weaving, endlessly famous in political literature," is thus presented, Foucault argues, as an *alternative* to the model of the shepherd (*STP* 145/149). It allows Plato to distinguish the art of the politician strictly speaking from all the related or auxiliary arts that are necessary to the exercise of the political but are not an essential part of it. Just as the weaver has certain requirements or preconditions without which he cannot weave (wool that has been sheared, yarn that has been twisted and carded, and so on), so the statesman needs certain elements or conditions without which he cannot be a statesman (generals ready for war, legislators set to make law, tribunals fit to pronounce judgment, and so on). In neither case, however, should one confuse these arts with the elements that make them possible (303d–305e). Hence "political action in the strict sense, the essence of the political," says Foucault, consists in joining together these various elements, just as—following the analogy—"the weaver joins the warp and the weft." Here is how Foucault describes this *essence* of the political:

> The politician will bind the elements together, the good elements formed by education; he will bind together the virtues in their different forms, which are distinct from and sometimes opposed to each other; he will weave and bind together different contrasting temperaments, such as, for example, spirited and moderate men; and he will weave them together thanks to the shuttle of a shared common opinion. So the royal art is not at all that of the shepherd, but the art of the weaver, which is an art that consists in bringing together these lives "in a community that rests on concord and friendship" (311b). In this way, with his specific art, very different from all the others, the political weaver forms the most magnificent fabric and "the entire population of the state, both slaves and free men," Plato goes on to say, "are enveloped in the folds of this magnificent fabric." In this way we are led to all the happiness a state is capable of. (*STP* 146/149–150)

Like the weaver, then, the statesman takes what is given over to him by the cooperative or auxiliary arts in order to weave together the various elements and characters of the polis into a harmonious whole. Like the weaver who brings together woof and warp, the statesman combines opposing virtues in order to achieve the right balance of character in the state.

We thus come to see at the end of the *Statesman*, Foucault says, that the royal art is the art of the weaver, not the shepherd. The *Statesman* would thus be Plato's "bona fide rebuttal of the theme of the pastorate" (*STP* 146/150). While the theme is not "entirely eliminated or abolished," it is restricted, as we saw Foucault argue earlier, to "minor activities that are no doubt necessary for the city-state, but that are subordinate with respect to the political order, such as the activities of, for example, the doctor, the farmer, the gymnast, and the teacher" (*STP* 146/150). All of these other activities might rightly be compared to that of the shepherd, but not the activity of the statesman, which has been definitively distinguished from all these others.

Foucault goes on to distinguish the art of the statesman from all other arts by arguing that only the practitioner of the former distances himself, like a weaver rather a shepherd, from those over whom he rules or exercises care. Definitive proof of this comes, Foucault claims, at *Statesman*

295a, where the Stranger says: "How could anyone, Socrates, sit beside each person all his life and tell him exactly what is proper for him to do?" (295a–b), a line that Foucault paraphrases in this way: "Do you think that the politician could lower himself, could quite simply have the time, to act like the shepherd, or like the doctor, the teacher, or the gymnast, and sit down every citizen to advise, feed, and look after him?" (*STP* 146/150). Foucault's parsing of Plato is telling, as well as deceptive, for while it sounds as if Plato is here *contrasting* the politician to the practitioners of these other arts, everything both before and after this line suggests a *continuity* or *comparison* between them. In the Age of Zeus, it is not only the politician or lawmaker who cannot sit down to advise everyone all at once and over time but the doctor and trainer as well. That is why these latter, just as much as the former, must issue general prescriptions for the majority of men. It is why they, just like the statesman, must *supplement* their very specifically tailored and timely advice with general prescriptions that are to be repeated and passed on through either oral or written law. What Foucault thinks is impossible for the statesman but possible for the doctor or the trainer is impossible for these latter as well. Because it is impossible to be present to all one's patients at once and for all time, general prescriptions—spoken or written laws of some kind—are necessary to supplement the living presence of the doctor or the trainer. In the Age of Zeus, *no one* can really be compared to the divine shepherd, who overcomes the limits of time and space, the limits, in short, of human finitude, in order to be present to everyone at once throughout all time. The problem with the statesman's art is therefore not unique to statesmanship.

Undeterred by any of these potential objections, intent, it would seem, on explaining away what appears to be a striking exception to his thesis regarding the introduction and importance of pastoral power in the West, Foucault concludes:

> Above all, let's not say that the politician is a shepherd (*berger*). The royal art of prescribing cannot be defined on the basis of pastorship. The demands of pastorship are too trifling to be suitable for a king. It is too little also because of the very humbleness of its task, and consequently the Pythagoreans are deceived in wanting to emphasize the pastoral form, which may really function in small religious and pedagogical communities; they are wrong in wanting to emphasize it at the level of the whole city-state. The king is not a shepherd (*pasteur*). (*STP* 147/150)

According to Foucault, then, Plato would have adopted this theme of the pastorate from the Pythagoreans, but instead of valuing it, as it might appear, he criticized it—at least in the *Statesman* where the more appropriate model of the statesman as weaver replaces the inferior model of the statesman as shepherd. When we combine "all the negative signs given by the absence of the theme of the shepherd in classical Greek political vocabulary" with "the explicit criticism of the theme by Plato," we are led to conclude, says Foucault, that "Greek reflection on politics . . . excludes this positive valuation of the theme of the shepherd" (*STP* 147/150).

Imitation and the Model of the Statesman as Shepherd

On a first hearing, and without reading Plato's dialogues alongside Foucault's analysis of them, Foucault's thesis sounds somewhat plausible and evidence from the dialogues is marshaled forth to support it. But because Foucault must argue not that the theme or model of pastoral power is simply not *present* in Plato, an argument that would be immediately refuted by the *Republic* and, especially, the *Statesman*, but that Plato is himself criticizing this theme or model even in those dialogues where he *seems* to be defending it, a more complete reading of the dialogues in question is not only legitimate but necessary. Foucault has argued, for example, through a dubious paraphrase of Plato, that while the king cannot be compared to a shepherd, the doctor or trainer can. But insofar as the statesman's art is the only one that oversees all the other arts, insofar as the statesman is the only one who has the good of the whole polis in view, it might be argued that only the statesman can really lay claim to being a "shepherd" of the state. Indeed it might even be said—and we here return to the theme of imitation—that the statesman is the only one capable in the Age of Zeus of *imitating* the divine shepherd in the Age of Kronos, the only one capable of *recollecting* through a kind of *political anamnesis* that lost age.

These themes of imitation and recollection are so central to the *Statesman* that is it hard to imagine a coherent reading of the dialogue that does not take them into account.[10] Moreover, the theme of imitation first emerges in exactly the place where, in the myth of the two ages, politics emerges, namely, in the transition from the Age of Kronos to that of Zeus. As we saw in the previous chapter, it is right after the divine shepherd has taken his hand off the universe, causing it to turn in the opposite direction of its own accord, that everything—including man—began "imitating" the universe, with each part having to rule over itself rather than being ruled over by the deity (273e). Just a page later, the Stranger repeats this thought, saying that it was by "imitating" the universe that humans came to make use of the arts in order to "direct their own lives and take care of themselves" (274d). This is the first kind of "imitation" in the *Statesman*, imitation by analogy or homology, as everything at the beginning of the Age of Zeus begins "imitating" the universe's self-rule in the absence of a guiding divinity.

The second use of imitation, somewhat later in the dialogue, occurs after the Stranger has argued that the true statesman, like the true physician, is the one who rules or cures only "by art or science" with no other consideration in view than the health or preservation of the state or the patient (293b)—a definition, note, that is perfectly consistent with Socrates's understanding of the true shepherd in Book 1 of the *Republic*.[11] According to the Stranger, this is the only "right definition" (*horon orthon*) of the statesman, the only true form of government, and all other forms must be considered not legitimate or even as really existing but as "imitating" (*memimēmenas*) this one true form (293c–e). Moreover, the better one imitates this one true form, the better the state is governed.

As for the one true form that is imitated, the analogy the Stranger uses is much more resonant with the model of the statesman as a shepherd rather a weaver. So long as the statesman is acting with "science and justice" (*epistēmēi kai tōi dikaiōi*) to preserve and benefit the state—a claim we will take up in the next chapter—it does not matter, says the Stranger with a striking euphemism, whether he puts the city on a diet by killing or banishing or sending some of its citizens out to form colonies or whether he allows it to grow by bringing in citizens from elsewhere (293d–e). This characterization of the statesman—which occurs, it should be noted, well *after* the model of the statesman as weaver has been introduced in the *Statesman*—recalls the way in which the shepherd in the *Laws*, often thought to be Plato's last dialogue, must sometimes "purge" his flock of a few of its members for the sake of the flock as a whole (*Laws* 735b).[12] It is, let me underscore, the image of the city as a flock and the ruler as a shepherd—and not as a weaver or a navigator—that facilitates this thinking of a purge or a purification of the city. Foucault's claim that the *Statesman* is a bona fide rebuttal of the model of ruler as shepherd meets here yet another very high hurdle to clear.

There is thus a relation of *imitation* between the Age of Kronos and the Age of Zeus, just as there will be, as we will see in greater detail in Chapter 6, a relation between the one true form of government and the six imitations of that form. How this imitation takes place is, of course, not obvious (indeed this is the whole enigma—the whole promise and problem—of Platonism), but it is central to Plato's understanding of statesmanship. The relationship between the ages of Zeus and Kronos is thus, clearly, much more complicated than Foucault has suggested. Somehow or other the many statesmen in the Age of Zeus are supposed to imitate the one true statesman in the Age of Kronos, law is somehow to imitate a state before law, the art of politics is supposed to bring about a state that precedes the political. It is only appropriate that an image of the statesman from the city (the weaver) would be used to imitate, approximate, and so bring about as far as this is possible a statesman whose image of rule comes before the state (the shepherd). The true or genuine weaver will thus be— or will have been—a shepherd. The image of the weaver will not replace but, on Plato's account, *supplement* the image of the shepherd. The statesman as weaver plies his art in such a way as to return, to the extent this is possible, to a time before art, to a time before the city and before politics, to a time when the "statesman" was a "shepherd." The supplement of the arts and of culture aim to return to—to recollect—an even more original, more natural state of affairs. The image of the shepherd, like that of the navigator, is, therefore, never simply abandoned by Plato in favor of the image of the weaver but integrated into a much more complex dialogical weave. Rather than speak of the rejection or refutation of the model of the statesman as shepherd, it would be more accurate to speak of the incorporation or transformation, the sublation, even, of that model.

Simply to say, as Foucault does, that the human shepherd is too much like other men to be a shepherd is to write off without explanation the entire philosophy of Plato; for it could then be said with equal merit that the particulars are too particular ever to recall a form, that the ideal is too much of an ideal ever to be imitated, participated in, or recollected. Whatever one thinks about this Platonic project, a reading of the *Statesman* and other related dialogues must at the very least attempt to take it into account. Were imitation a completely subordinate or insignificant notion in Plato, then Foucault could be forgiven for neglecting it. But because it is, in some dialogues, another way to rethink participation and recollection, the relationship between particulars and form, examples and essence, the many and the one, becoming and being, it is absolutely central to Plato's philosophy as whole and, it would seem, to the *relationship*, which is surely not simple, between the many, imperfect statesmen in the Age of Zeus and the one divine, ideal ruler in the Age of Kronos, or else, more pertinently still, between that part of the polis (or that part of the statesman, or of the citizen) that can be likened to a human being and that part that can be likened to a god.

It is in just these terms that an important passage from the *Laws* seems to echo, reformulate, and clarify the myth of the *Statesman* that we have been following in this chapter. In this retelling of the myth of the two ages it becomes clear that the best polity is not simply the one that existed in some golden age of the past but the one that exists whenever the divine rules *in us*. There once existed in the time of Kronos, says the Athenian, a "most prosperous government and settlement, on which the best of the states now existing is modeled [*mimēma*]" (*Laws* 713b), the suggestion being once again that this mythical past regime is never simply left behind or overcome but is *imitated* in the present.[13] As in the *Statesman*, life was blissful during this earlier age, with all things being furnished by "spontaneous growth [*automata*]." A nobler and more divine race of *daimons* is again said to have watched over mankind during this earlier age, just as, in an analogy we could almost predict by now, mankind today rules over oxen and goats (713d). Kronos ruled over men in this age, we are told, out of his "love of humanity"—that is, out of his *philanthropy*, a distinguishing characteristic not only of the Pythagorean ruler, as Foucault points out, but, as we suggested earlier, of the Platonic Socrates.

This account of the Age of Kronos is thus in almost complete conformity with the *Statesman*. But then there's a detail that adds something new to the *Statesman*, a detail that helps explain much of what goes unexplained but is probably implied in this other, no doubt earlier dialogue. The Athenian says that his story teaches us that a state is never free from ills when a man and not a god is ruling over it and that we must therefore "imitate" (*mimeisthai*) the life of the Age of Kronos (*Laws* 713e). We do this, he says, by ordering our houses and our states in "obedience to the immortal element within us, giving to reason's ordering the name of 'law' [*tēn tou nou dianomēn eponomazontas nomon*]" (*Laws* 714a). Law (*nomos*) in conformity with reason (*nous*)—that, says the Athenian, is the immortal element within us, and only states that are ruled on the basis of it will prosper.[14]

Hence the Age of Kronos functions even more clearly here in the *Laws* than in the *Statesman* as a sort of mythological or analogical account of what takes place "within us."[15] It is not too much to say that, for the Plato of the *Laws* and, arguably, the *Statesman*, the kingdom of Kronos is within, and we must do what we can in the Age of Zeus to bring it about both within us, within our souls, by imitating and allowing ourselves to be ruled by divine reason and, perhaps, outside us, in an actual state, should an opportunity for this arise. There is, says the Athenian, no salvation for a man or a state that is ruled by something other than this law of reason (*Laws* 714a–b), no salvation, it might be said, for a man who does not allow the immortal part of his soul to watch over the other parts like a shepherd tending to his flock, and no salvation for a state that does not have leaders who attempt to imitate through their science and law the divine shepherd in the Age of Kronos. We thus need to think about the two ages as happening not one after another, successively, but, as it were, simultaneously, the one intersecting or imitating the other, just as the circles of the Same and the Other will move simultaneously in different directions in the *Timaeus* (36b–d).[16]

In the *Critias*, too, myth is used to suggest that the past is not simply something to be memorialized but an ideal to be recollected through a kind of *political anamnesis*. Critias there describes not only the lost city of Atlantis, its geography and its government, but the ancient city of Athens, where the gods had divided up the land by lots, with each god settling in his or her own region, gods who then "reared us up, even as herdsmen rear their flocks, to be their cattle and nurslings" (*Critias* 109b). Hence the city of Athens was once divided up into various regions ruled over by various gods, various divine herdsmen, as it were, much like the universe during the Age of Kronos. But, says the *Critias*, introducing a new element to what we see in the *Statesman*, or rather, "internalizing" what we have seen there, "it was not our bodies that they constrained by bodily force, like shepherds guiding their flocks with stroke of staff, but they directed from the stern where the living creature is easiest to turn about, laying hold on the soul by persuasion, as by a rudder [*oiaki*], according to their own disposition; and thus they drove and steered all the mortal kind" (109c). This word *oiax*, "rudder," is the same one used in the passage in the *Statesman* where the divinity during the Age of Kronos is said to rule or steer the universe. In the *Critias*, this rudder is "persuasion," a kind of middle term, as always in the Greeks, between force and reason. We thus see Plato using here in a single passage *both* the model of the statesman as shepherd *and* the model of the statesman as navigator, further evidence that none of these tropes are ever definitively abandoned but are always being redeployed according to the dictates of the argument or the dialogue.[17]

Critias goes on to say that the Egyptians had preserved the names of famous Athenian men and women from this period and he describes a social organization that resembles much of what is recounted in the *Republic* (*Critias* 110b): Men and women share in all tasks, the military class dwells apart and without private property, all things are considered to be common property, and so on (*Critias* 110c). Hence we see that the state sketched out in speech by Socrates in the *Republic* has a historical or quasi-historical counterpart, as if the task of seeing such a state in motion, to paraphrase Socrates from the *Timaeus*, were less a matter of inventing a new state or implementing the city in speech of the *Republic* than of recollecting and imitating through a sort of political anamnesis the glorious past of Athens. The city that Athens must recollect is thus none other than itself, a past it once lived (at least in myth) but now can no longer remember. In order to recall its own past, then, what is required is the *supplement* of another's archive. Only through the written memory of the Egyptians is Athens able to recollect its past—a past that resembles in many respects life during the Age of Kronos. Hence the living memory of Athens needs the dead archives of Egypt in order to recall its own past, an age, we are to infer, that must be recollected and thus imitated in order to bring about the best possible state in the present.

Foucault argues that, in the *Statesman*, Plato evokes the model of the shepherd-king in order to *reject* it or, rather, in order to *replace* it with the image of the statesman as weaver. But as we have just seen, it is not Plato's habit to move from one image or analogy to another in this way, replacing a less appropriate image by a more appropriate one. Plato is able to accommodate many different images and analogies: if one comes after the other, this does not mean that the first is being replaced by the second.[18] A good case in point, as I suggested earlier, is the image of the statesman as navigator of a ship. Foucault himself mentions this image in *Security, Territory, Population*, but he does not mention the fact that Plato uses it in several dialogues, from *Republic* and the *Laws* to, precisely, the *Statesman*, where the divine ruler in the Age of Kronos is compared to the navigator of a ship who keeps his hand on the rudder or tiller of the universe until it is time for him to release it and let the universe move of its own accord. The image of the statesman as weaver will not disqualify the image of him as a navigator any more than it will disqualify the image of the statesman as shepherd or, indeed, as physician.

If the theme of the ruler as shepherd is definitively refuted, as Foucault believes, in the *Statesman*, then we must somehow explain its return in dialogues that are generally acknowledged to come after the *Statesman*, for example, *Critias* and, especially, *Laws*. It appears instead that the theme or the model of the ruler as shepherd proved useful to Plato in ways or places that the image of statesman as weaver did not. When it came, for example, to portraying the statesman as one who must look out for the whole of the *polis*, who must look out for its health, who on occasion might even have to administer a purge, who, in short, is responsible for the *salvation* of the *polis*—this metaphorics of salvation being central to both the *Statesman* (see 297b) and, especially, the *Laws* (see 714a–b, 715c)—the image of the weaver appears much less appropriate than that of the navigator or the shepherd.

The statesman is indeed sometimes in Plato still a shepherd. Were Foucault to have admitted this, he would have been able to acknowledge the enormous exception Plato is with regard to classical political Greek thought. Such an acknowledgement would have also given him the chance to take into account the genuine *danger* presented by this political thought, the danger that is to be found in any political discourse that tries to return to an earlier, mythological age, that tries to *recollect* a more natural political state in the name of *saving* the state. Without giving an account of precisely how this political thought combines pastoral power with a discourse of salvation, one will never be able to understand why—to reverse the famous words of the poet and of the philosopher who cites him—wherever the saving power is, there grows the danger.

As we will see, it is through a kind of recollection or anamnesis that the individual recovers the divine or immortal part within him, and it is through a kind of political anamnesis that the state and its leaders are able to bring about something of the divine in the polis. The true statesman might thus still be said to be the only *true shepherd* in the polis, the only one capable in the Age of Zeus of imitating the divine shepherd in the Age of Kronos, the only one in the Age of Zeus capable of *recollecting* that lost Age of Kronos, the only one in *this life*, in what we *call* life, who is able to imitate that *true life* that is promised in that earlier, mythical age.

CHAPTER 4

The Measure of Life and Logos

... the old word *vie* [life] perhaps remains the enigma of the political
around which we endlessly turn.

—JACQUES DERRIDA, *Rogues*

In the first chapter, we saw the Stranger use a series of terms that suggested a kind of organicism in the things themselves or in a conceptual landscape that needed to be divided or carved up along its natural joints.[1] By following the term *diairesis* back to the *Phaedrus*, we then saw that a good speech must also have joints, that a good *logos* must be organized like a *zōon*—that is, like an ensouled being. It was as if a certain conception of life had insinuated itself into both being and *logos*, as if the first line to be drawn— but by whom or what?—were the line between things and words, being and *logos*, while "life itself" slipped away so as to infiltrate or animate both sides of the division.

In the following chapter, we focused on a strange moment in the myth of the two ages where the word *automatos* seems in its five uses to remain ambivalent or undecidable, suggesting at once or in turns a kind of spontaneous life, a way of describing an interior principle of self-motion for that *zōon* that is the universe, and the lack of any motivating or guiding principle, a universe that, at the withdrawal of the Demiurge, is cut off, set adrift, and left to degeneration and dissolution. At the beginning of the Age of Zeus, the universe seems to turn of its own accord, but this also means— and means perhaps first and foremost—that it turns *without* the orientation and good guidance of the Demiurge, who is described not only as a shepherd or a helmsman but a Teacher and a Father. Were it not for the possibility of *recalling* those teachings, the universe would be as lost or bereft as the disciples of Heraclitus, who were reduced to living *automatos* in the absence of any teacher. Were it then not for the periodic return of this Teacher and Father, the universe, wandering like an orphan, would sink into the endless sea of unlikeness, all order overcome by diversity and all identity undone by difference.

We thus begin to see a configuration develop whereby the Father/ Teacher/Demiurge not only orients and directs all the *zōa* under his care and guidance but also provides them with *life*, not this corruptible and death-bound life that we *call* life but something like a true life or life itself. Life, so-called life, in the Age of Zeus would thus deserve the name life only to the extent that it imitates or recalls life itself—that is, only to the extent that one lives a life in accordance with that ideal life that no living human will have ever actually lived. There would be, then, on the one hand, life, true life, the Father as giver of life, the Father as a Teacher whose *logoi* are like living beings, and, on the other, the withdrawal or absence of the Father, the foundering of the universe in his absence, and the necessity of staving off the worst in the absence of the best by means of the arts, self-governance, and laws. That leads me to the topic of the present chapter: the *Statesman's* treatment of *written law* as what is *second best* in the absence of a living Father, Teacher, or Lawgiver. My hypothesis will be that the myth of the two ages gives us a model for thinking the relationship between science and law, and particularly written law, in terms of what we have already seen as two different ways of thinking *life*. That is where we are headed, but in order to get there we need to look at some of the sections of the dialogue that lead up to the Stranger's treatment of written law.

Conclusions of the Myth

The Stranger argues that the myth of the two ages reveals two problems with their previous argument and division (274e–277c). First, they combined and so confused the divine and heavenly shepherd with human rulers by grouping them both under the name "herdsman," even though only the former is truly a herdsman or shepherd different than the herd. Second, they did not distinguish the manner or mode of rule in order to contrast two types of care-taking—the voluntary and the compulsory, the one exercised by the king and the other by the tyrant. Let's look at these two mistakes in a bit more detail, especially since they seem to suggest that Plato is indeed, just as Foucault suggests, *criticizing* the model of the human ruler as shepherd.

The myth was introduced, says the Stranger, to show that the only one who deserves the name "shepherd of the human flock" is the demiurge in the Age of Kronos. In other words, the "form [*schēma*] of the divine shepherd is greater than that of the king" inasmuch as statesmen in the Age of Zeus are by nature much more like their subjects, having had the same human upbringing and education as they did. The fact that the king is not so different from his subjects also explains why so many other men can lay claim to taking care of the flock. No one makes a similar claim with regard to the divine shepherd, who is at once superior in nature to his flock and able to do everything for his flock. (We are not far from one of the questions that haunts the *Republic*, the question of how to educate the educators, for it would seem that only an educator born outside the system could be free enough from the corruptions of human education to become a true shepherd or statesman.)

The Stranger argues that instead of defining the human shepherd as feeding his flock, they should have just described him as *caring* for it. The human shepherd—the statesman—does indeed *care for* his people but he does not actually *feed* them as the divine shepherd does. Indeed, says the Stranger, "no other art would advance a stronger claim than that of kingship to be the art of caring [*epimeleia*] for the whole human community and ruling all mankind" (276b). With a more general category like *caring*, says the Stranger, they can "wrap up the statesman with the rest" and then divide as they did before (275e): caring for winged as opposed to land animals, those that have horns as opposed to those who do not, those who mix breeds as opposed to those who do not, the four-footed as opposed to the two-footed. In this way, the definition of *caring* will cover kingship in the Age of Kronos as well as the Age of Zeus. It is not nourishment but care (*epimeleia*) that is central to the statesman's art, the very same notion of care that is so central to many of Plato's dialogues and, interestingly—and, particularly in the form of self-care or care of the self (*souci de soi*)—to Foucault's analyses of ancient texts in *The History of Sexuality* and elsewhere.

Conflating the human with the divine ruler by suggesting that both do everything for their flock was thus their first mistake. Their second mistake was in not distinguishing the "modes" of ruling, which caused them to lump together the king and the tyrant. After having distinguished the "divine shepherd" and the "human caretaker," they could have gone on to distinguish within this latter "compulsory and voluntary" (*tōi biaiōi te kai hekousiōi*) caretaking (277d). For the tyrannical, which requires compulsion, is much different, says the Stranger, than the voluntary, which is properly "political" and is what "the true king and statesman" possesses. In the Age of Kronos, rule is, properly speaking, *neither* compulsory *nor* voluntary, though it could be said negatively that the divine shepherd, without ruling by force, nonetheless rules over his subjects without their consent. Only in the Age of Zeus, then, can we distinguish compulsory rule or rule by force from voluntary rule and, as a result, the tyrant from the king or the statesman. The difference between compulsory and voluntary rule over subjects would thus seem to distinguish tyranny from the art or science of statesmanship. The problem with this claim, as we will see, is that the status of statesmanship as an art or science will have played no real role in this distinction.

Learning by Example: The Example of Letters

Oblivious to any of these problems, the Young Socrates says, once again, that he is satisfied with the results of their search. The Stranger reminds him, however, that one of the rules of this sort of dialogue is that *both* interlocutors must agree and that he himself is not yet satisfied. As he puts it, their "figure [*schēma*] of the king is not yet perfect," for they are like sculptors who, "in their misapplied [*para kairon*]" enthusiasm—that is, in their inopportune enthusiasm—have made too many and too large additions, thereby making their work take much longer (277a). They used "illustrations" (*paradeigmata*) to explain the king, and a "marvelous mass of myth," and so have used more than they should have, making their discourse too long without yet really depicting the statesman. Shifting from an analogy regarding sculpture and the proper size and proportions of a discourse to one of painting and the requisite clarity, color, and detail of a discourse, the Stranger says that all their talk, "just like a picture of a living creature [*ho logos hēmin hōsper zōon*], seems to have a good enough outline [*tēn exōthen men perigraphēn*]," but it has "not yet received the clearness [*enargeian*] that comes from pigments and the blending of colors" (277c). Once again, *logos* is compared to a *zōon*, just as it is, as we saw in Chapter 1, in the *Phaedrus*, and just as it is in the *Timaeus*, as Socrates, near the beginning of that dialogue, likens his discourse of the previous day—that is, his description of a state that resembles in many ways the one laid out by him in the *Republic*—to a living being, a *zōon* (whether a real *zōon*, a real animal, or else, because *zōon* can mean not only a living being but a picture, the picture of a *zōon*, the *zōon* of a *zōon*) that he would like to see in motion.

Hence the Stranger compares their *logos* to a *zōon*—that is, to the picture of a *zōon*. Their discourse, which must itself be organized like a *zōon*, has thus far provided only an *outline* of the animal or living being, the *zōon*, they are pursuing, namely, the statesman, an outline that must now be made clear and vibrant, filled in with living colors. The Stranger goes on to contrast, more or less in passing, it seems, though the distinction is crucial, the relative merits of portraying a living thing, a *zōon*, by means of discourse, *logos*, on the one hand, or painting or sculpture on the other: "it is more fitting [*prepei*] to portray any living being [*pan zōon*] by speech and argument [*lexei kai logōi*] than by painting or by any handicraft [*graphēs de kai sympasēs cheirourgias*] whatsoever to persons who are able to follow argument; but to others it is better to do it by means of works of craftsmanship" (277d). Speech or *logos*, a verbal depiction or account, is more fitting or appropriate than painting or sculpture or any handicraft for depicting a *zōon*, claims the Stranger, insofar as only speech is or should be itself organized *like* a *zōon*, like an animal or an organism. Speech is what is most proper for depicting such things as the nature of the king, so long, that is, as the audience is able to follow. For those ill-suited to follow a discourse or *logos*, a statue or picture will have to do; that is, it will have to stand in for or supplement a properly verbal depiction. Once again, just as we saw in Chapter 1, we have the opposition between the work of the head or of the mouth, *logos*, and that of the hand, handiwork, and so speech as opposed to painting and sculpture and, perhaps, as we will soon see, to writing. Though only one *zōon*, namely *anthrōpos*, has *logos*, *logos* would be the proper measure of all *zōa* and the proper means of depicting them.[2]

Discourse, *logos*, must be organized like a *zōon*, like a living being, in a way that would fit or befit, that would be appropriate to, the nature of the thing or idea being depicted, which is also, as we have seen, organized like a *zōon*. There would be, then, a fit, a *homology*, between the saying and the said, the discourse and its subject, both of which would be or should be organized like a *zōon*, like a living being. As in *Timaeus* (29b–d), where discourses or accounts can only be as good as the objects for which they are accounts, where only discourses regarding eternal, certain objects of knowledge can themselves be certain, while discourses of things in the sensible realm can be only probable, so, by a similar principle of like to like, one that Plato and so much of the Greek world will have inherited from Homer, only a discourse ordered like a living being is appropriate to depict that which is organized like a living being. Both being *and* logos would thus seem to be related to life—as if life, whether *bios* or *zōē*, we will have to see—were the third term that underlies the two, as if the first line to be drawn, as I suggested at the outset of this work, were the line between being and logos, a line that is perhaps drawn with and within life itself, or perhaps even *by* life itself or, rather, by what I suggested at the outset should be called *life-death*.

To explain, then, where their account is lacking, the Stranger says he will give an example (a *paradeigma*) (277d–278e). For in order to set forth greater ideas, he explains, it is necessary to give examples: "for it would seem that each of us knows everything that he knows as if in a dream and then again, when he is as it were awake, knows nothing of it all" (277d).[3] By teaching us to see the relations between things in different contexts, examples help us transfer our ideas from one realm to the next—that is, they help us move from the dream state of right opinion to wakefulness and knowledge, from a death-like state to a life-like one. As the Stranger will say a bit later, they must first "see in another small and partial example the nature of the example in general" in order to help them to think the nature of the king—that is, "in order that this may be waking knowledge for us, not dream knowledge" (278e). We thus begin to see more clearly what Plato has in mind when he contrasts dream knowledge with waking knowledge. Waking knowledge is a knowledge of things as they are, not just as they appear, a knowledge that sees just how things are examples of their idea; dream knowledge is knowledge (or, better, right opinion) about the particular alone, where one has a true opinion about the particular example but no knowledge of that of which it is an example— that is, no true understanding of the self-same thing in which the particular participates. It is perhaps too much to compare this waking knowledge to the Age of Kronos, a knowledge that is awakened by means of right opinion in the Age of Zeus, a right opinion that will be compared, in just a moment, to writing, but we are already beginning to see the outline of such a configuration beginning to take shape.

The Stranger proceeds to teach Young Socrates about the need for examples in learning by giving him an example of examples, a paradigm for understanding paradigms. In other words, by using an appropriate example or paradigm, the Stranger will be able to show the Young Socrates how examples or paradigms work in general. The Stranger recognizes the odd or absurd nature, the regressive or recursive nature, of this approach: "I seem at present in absurd fashion to have touched upon our experience in regard to knowledge [*epistēmēs*]," for "the very example I employ requires another example" (277d). To explain why examples are necessary to understanding, to explain how examples work, the Stranger needs to give an example of examples.

In order to show where their discourse is defective, the Stranger argues that an example is necessary. This then leads to an explanation, by means of an example, of the necessity of learning through examples. The example of examples that is then proposed turns out to be the example of letters. When children are learning letters, says the Stranger, they are able to "make correct statements [*talēthē*]" about them in certain short syllables but "err in opinion and speech [*doxēi te pseudontai kai logōi*]" about these same letters when they are in other syllables (278a). The best way to help children learn, then, is to juxtapose those cases where they have "correct opinions" (*orthōs edoxazon*) about letters—and not knowledge, notice, since knowledge requires being able to see the sameness of a letter beyond each particular example—with those cases where they do not yet recognize the letters. The point is to show that the "nature" (*physin*)—that is, the being, essence, or identity—of the letter is the "same" in both combinations. The letter is thus seen to be the same "either by differentiation from the other letters, in case it is different, or because it is the same" (278b–c). Children learn when a letter they know in a simple syllable is juxtaposed with its use in a more complex syllable that they do not yet know. Hence "an example is formed when that which is the same in some second unconnected thing is rightly conceived and compared with the first, so that the two together form one true idea [*mian alēthē doxan*]" (278c). In other words, an example mediates between what we know and what we do not. Learning letters by means of example is thus an example of how we learn by example.

Two further points about this particularly rich passage. First, the one who gives an example, not unlike the one who uses *diairesis*, must already see in advance where he is going—that is, he must see that of which his example is an example. The teacher chooses an example of a letter because he knows that by means of that particular example the student he is trying to teach will learn. Examples work, in other words, only when someone who already sees a relation—and who sees that someone else does not—points out that relation to that someone. Giving an example, an appropriate example, thus requires a kind of knowledge, a pre-understanding or foresight, of that of which the example is an example and of the kind of example that will be appropriate for the one who does not yet know. As Homer says and Plato loves to repeat, "when two go one sees before the other" (see *Alcibiades II* 139e; citing *Iliad* 10.224). It is the one who "sees before" who will help the other see by means of an example. Hence an example is always *oriented*, to return to the word that guided us in the previous chapter, by the pedagogical purpose that informs it.[4] Plato's criticism of Heraclitus, we recall, was not just that neither he nor his disciples could see but that neither he nor his disciples could help others to see. Neither, in short, would have been able to use an appropriate example to help someone else understand. As for the Young Socrates, he is, as the Stranger knows, in need of examples in order to learn what the Stranger wishes to teach him. The Stranger—and on another level Plato—has thus chosen the example of writing and, even before that, the example of examples, as a way of orienting, on the one hand, the Young Socrates and, on the other, the reader more generally in the search for the nature of the statesman.

An example is thus always oriented, always offered in view of some goal by someone who knows or thinks he knows. And yet, and this is my second point, every example also always exceeds its purpose; that is, it always says more than it purports to say. The example of writing, for example, is thus obviously not just one example among others in a dialogue where the written law will become a central theme, the necessity of written law in the absence of the lawmaker or ruler.[5] In the example of learning letters, we have not only an example of how we learn by example but an example that speaks already about writing, sameness and difference, technicity and nature, and perhaps the human statesman in the Age of Zeus and the divine herdsman in the Age of Kronos. While speech, too, no doubt requires such sameness and difference, Plato's example of the example of writing says more than it says about the eventual evaluation of writing not only later in the *Statesman* but in the *Phaedrus* and elsewhere. (It is, in effect, by weaving together a series of similarities and differences, or presences and absences, in letters that one first learns to read words, and by weaving together nouns with verbs that one learns how to make and to read propositions or *logoi*.[6]) All this is another way of saying that no example is ever a *mere* example, that no example is ever completely neutral or innocent. And this is the case even when—perhaps especially when—one believes oneself to be in complete control of one's example for some pedagogical purpose.

Learning how to read by means of appropriate examples of letters is thus a good and appropriate example of how people learn by example in general; in other words, it is a good example of how we learn *ideas*. Just as children have a hard time moving from letters in simple syllables to the same letters in more difficult ones, so our soul sometimes has difficulty, says the Stranger, moving from simple to more difficult combinations of "letters or elements [*stoicheia*]"—the term *stoicheion*, meaning here both *letter* and *element*, yet another indication of the homology between being and logos. While the soul is sometimes "firmly grounded in the truth about every detail," in other cases "it is at sea about everything"; while it sometimes has "correct opinions" about certain combinations, it is ignorant "of the same things when they are transferred to the long and difficult syllables of life" (278c–d). The Stranger is thus transferring here the example of letters, which first shows the necessity of examples and then shows how examples work, to learning ideas more generally. Again, without knowledge or at least correct opinion, with just false opinion, there would be no way to offer appropriate examples so as to help another, as the Stranger says, "attain to even a small part of truth and acquire wisdom" (278d). Once again, we need some guide, someone who can "see" in advance, who has some "foresight." Without such a guide, we risk learning nothing; with him, we can begin with a single truth and be led, as the *Meno* puts it, to "discover everything else" (*Meno* 81c).

The Example of Weaving

Having shown the necessity of learning by example, which itself required an example of learning by example, namely, the example of learning letters, the Stranger can now introduce an example that, he believes, will help the Young Socrates and, a fortiori, the reader or listener to learn about the nature of statesmanship. That is the example of weaving (279a–283b). It is this example, we are supposed to think the Stranger can foresee, that will be the right example to help Young Socrates to learn about the nature of statesmanship, the right example to help him, in a first moment, to learn how to separate off the statesman from others who might be confused with him and then, in a second moment, to help him to understand the proper activity of the statesman.

To separate the statesman off from the pretenders, we first need to understand how to separate something off from the things that are associated with it. Because, as we have seen, we learn by example, we now need a good example to help us understand how to separate in this way. Weaving, the Stranger suggests, is a good example of an art that can be separated off from its kindred and associated arts. Of course, as we will see—and as we suspect the Stranger, as the one who gives this example, already sees—the example of weaving will be not only a good example of how to separate off one art from things associated with it but a good example for understanding the very activity of the statesman.

In order to isolate the king from all those who also claim to "care" for the state, they thus need an example of something small that "has the same activity as statesmanship": the example chosen by the Stranger is wool-weaving (*hyphantikē*)—an example that is chosen by the Stranger because he already *knows* that there is a similarity between weaving and statecraft and that this example will help him to *orient* the dialogue and *educate* the Young Socrates. By concentrating on the right part of this activity, they will, he says, arrive at "the illustration we desire" (279b). To isolate this activity, they must separate off weaving from its *kindred* arts and then from its *co-operative* arts (by distinguishing contingent from actual causes—that is, causes that provide the tools for producing something and those that actually do the producing). I will not review here all the various distinctions that are made in the *diairesis* of weaving that follows (279a–283b). Other commentators, such as Seth Benardete and Stanley Rosen, have already done this, and with great care and insight.[7] For me, the main point of this *diairesis* is to introduce weaving as the supplement—though not the replacement—to shepherding and as the image, or at least part of the image, of dialectic itself.[8] Near the very end of the *diairesis* separating off weaving from the arts related to it, the Stranger says—in words that should make us think of Socrates's definition of dialectic in the *Phaedrus*, though also of the very "art" of *diairesis* in which the Stranger is now involved—"in all things we found two great arts, that of composition and that of division [*hē synkritikē te kai diakritikē*]" (282b). At once mentioning and using, or noting and performing, these two great arts, the Stranger goes on to separate off the arts of division (the art of carding, for example) from the arts of composition, in order then to isolate the art of weaving among these latter. "Let us put aside," says the Stranger, "all that belongs to division, making two parts of wool-working, by applying the principles of division and composition" (282d). This latter can then be divided into the art of *twisting* threads and the art of *intertwining* them, a distinction that will eventually lead us to weaving, which will be defined as an art of composition or intertwining, the art of "joining together" (*symplokē*) threads. The Stranger concludes:

> when that part of the art of composition which is included in the art of
> weaving forms a web by the right intertwining [*euthyplokia*] of woof
> and warp, we call the entire web a woolen garment, and the art which
> directs this process we call weaving [*hyphantikēn*]. (283a)

As the Stranger suggested just a bit earlier, they have used the two great arts of dividing and combining, just as in Socrates's definition of dialectic in the *Phaedrus*, in order to separate off weaving as an art of combining— yet another indication of the way in which form and content, method and theme, discourse and object, are intertwined in this dialogue. Weaving as the art of combining threads has proven to be an appropriate example for showing how to separate off one art from another, an appropriate example for showing how dialectic—or at least a part of dialectic—works. Later in the dialogue, it will also be used as an image for statesmanship itself. The question will then be, once again, whether the image of the statesman as weaver actually replaces or, as I have argued, *supplements* the image of the statesman as shepherd—that is, whether weaving as an art of the city is not meant to *recollect* or *imitate* the "art" of shepherding that comes from out-side the city.[9]

Two Forms of Measure

At this point, the dialogue takes yet another turn, at once unanticipated and already foreshadowed (283b–284e). The Stranger asks whether it was necessary to make so many distinctions and draw the conversation out to such lengths in order simply to say that "weaving is the intertwining of warp and woof." The *diairesis* that resulted in a definition of weaving seems to have had even less significant results than the initial *diairesis* to define the statesman. Perhaps, as one says in French, *le jeu n'en valait pas la chandelle*—that is, the long discourse on weaving was not worth all the trouble. "Why did we beat around the bush and make a host of futile distinction [*en kyklōi pampolla diorizomenoi matēn*]," says the Stranger; that is, why did they go round in circles and draw things out without purpose (*matēn*) (283b).

To investigate, then, whether their discussion was too long—and so avoid in the future the "malady" (*vosēma*) of a discourse that is not of the proper length (as opposed, it seems, to a healthy discourse or *zōon* of the "right" length)—the Stranger proposes to "scrutinize the general nature of excess [*hyperbolēn*] and deficiency [*elleipsin*], for the sake of obtaining a rational basis [*kata logon*] for any praise or blame we may bestow upon excessive length or brevity in discussions of this kind" (283c). Hence they must now consider "length" and "shortness" and "excess" and "deficiency" in general—all subjects of the "art of measurement [*metrētikē*]" (283c).

Returning to the method of *diairesis*, which has been no more replaced than the model of the statesman as shepherd has been, the Stranger divides the science of measurement, into two parts, one concerned with "relative greatness and smallness" and the other with "something without which production would not be possible"—that is, the measure of the mean (*to metrion*) (283d). On the one hand, there are the arts that measure number, length, depth, breadth, and thickness in relation to their opposites and, on the other, those that measure these in relation to the "moderate [*to metrion*], the fitting [*to prepon*], the opportune [*ton kairon*], the needful [*to deon*], and all the other standards that are situated in the mean between the extremes" (284e).

The first of these arts of measure allows one to say that something is bigger or smaller than something else, bigger or smaller *relative* to something else. But we must also be able to say that something exceeds or is inferior to the mean, a mean that will depend on the context in which this something is found and the uses to which it is put. Only this second kind of measure makes possible any kind of moral or aesthetic or pragmatic judgments for assigning praise and blame in practical affairs of all kinds. Only this latter form of measure allows one, therefore, to praise a *logos*—or anything else—for being of the *right* length, or blame it for being *too* short or *too* long. Indeed, without this measure of the mean the arts themselves would be destroyed, it seems, everything from weaving to statesmanship, insofar as each of these arts regards excess and deficiency in relation to the mean "as real difficulties in actual practice [*peri tas praxeis*]." Only when these arts "preserve" the mean can it then be *said*, says the Stranger, that their works are "good and beautiful" (*agatha kai kala*), which means that ethical and aesthetic judgments in practical affairs all depend upon the measure of the mean (284a). It is a fascinating moment in an already infinitely fascinating dialogue, the first time Plato ever *thematizes* a form of measurement that he will have appealed to in almost every other dialogue before the *Statesman* without ever treating it as a separate class.[10]

The Stranger here recalls that just as they were forced to conclude earlier in the day in their search for the sophist that non-being exists, so they must now conclude that the measure of the mean exists, that greater and less are relative not only to one another but to the *mean*. All practical arts or efforts require the mean; indeed, without it "neither the statesman nor any man who has knowledge of *practical affairs* can be said without any doubt to exist" (284c). Hence a non-practical man might speak of founding a *polis* of five people or one of five million, but a practical man—even the practical man engaged in speculation, such as the Athenian of the *Laws*—must determine which of these *poleis* is too small and which too large for human habitation—that is, too small or too large in relation to a mean. To give another example, one might imagine a sculpture of Athena that is five centimeters tall and another that is five kilometers tall, but one must then rely upon the experience of the practiced sculptor to determine what size is most appropriate for *human perception*.[11] The measure of the mean is thus always a *human* measure, a measure that is related in one way or another to human utility or human judgment, as opposed to relative measure, which is always an abstract measure, a relation between two things rather than a measure that is attached to a particular situation that can elicit praise or blame from a human observer.[12] Unlike those purely theoretical arts, which can observe relations without concern for practical application, statesmanship requires the measure of the mean: One must know, for example, not just about the relative length or difficulty or costliness of wars but when it is opportune or proper to go to war and when not—that is, when war is too early or too late to pursue, when it is too costly and when it is not.

The question then becomes how this measure of the mean, this measure that is always related to man, fits with the rest of the dialogue and, in particular, with the myth of the two ages, which, as I have tried to argue, lays out not only many of the terms of the rest of the dialogue but its general schema. Recall that the Stranger and Young Socrates began their discussion of the two forms of measure by asking how they may bestow praise or blame on a *discourse* with regard to its length. Later in the dialogue, we will see that this second form of measure is really only used, indeed really only *can* be used, by those who are present to the situation to which they are responding. Only a physician who is present to his patient, for example, will be able to take advantage of the opportunity that presents itself in order to help heal him; only a lawmaker who is present to the polis will be able to respond in a way that is appropriate or proper or fitting to the changing situation of the polis. Rather than leaving written instructions to be followed in his absence, instructions that cannot respond appropriately to changing circumstances, the physician, with regard to his patient, and the lawmaker, with regard to the polis, is able to adjust his response to circumstances only if he is present to the situation—that is, of course, only if he is alive and able to respond with a living discourse to this live situation.

This should remind us, of course, of the difference between the Age of Kronos, in which God or the gods were our caretakers, able always to give us what is just right, and the Age of Zeus, where the God has withdrawn, leaving us with the task of approximating what is just right through law and the arts. The question then becomes whether these two forms of measurement, one of which seems to correspond to an Age of Kronos where the *kairos* is always found and attended to, correspond to two very different kinds of *life*.

The Proper Measure of Discourse

It is on the heels of this division of the arts of measurement into two classes that the Stranger makes another crucial methodological remark (285a–287a). He says that such division allows them to avoid lumping everything together into the same category, as most people tend to do, without difference or distinction. Here is how he describes what one "ought" to do—a description that ought to remind us of Socrates's praise of the arts of collection and division in the *Phaedrus*:

> when a person at first sees only the unity or common quality of many
> things, he must not give up until he sees all the differences in them, so far
> as they exist in classes [*en eidesi*]; and conversely, when all sorts of dissimi-
> larities are seen in a large number of objects he must find it impossible

to be discouraged or to stop until he has gathered into one circle of similarity all the things which are related [*ta oikeia entos mias homoiotētos*] to each other on the basis of their essential nature [*ousiai*]. (285b)

The reason for this methodological remark, as we soon see, is that the very object or goal of the dialogue may well be—in addition to defining the nature of the statesman—learning how to see similarities and dissimilarities in this way. In other words, the goal (or one of the goals) of the dialogue may well be to explain and to perfect, to mention and to use (and to improve one's skills by mentioning and using) the method itself. Returning to the previous object of their discussion—that is, to the example of the example raised earlier in order to demonstrate the necessity of learning through examples—the Stranger says that just as one asks children about letters not "for the sake [*heneka*]" of a particular word but so as to help children learn all the words in the lesson, so their investigation of the statesman has been undertaken not simply "for the sake [*heneka*]" of this one subject but in order to become "better thinkers about all subjects" (285d). As we saw, examples are always provided with some other end in view, and it is only the person who has already seen that "for the sake of which" something is being done who can provide the appropriate example. Just as the example of letters was introduced as an example of how we learn by example, an example of how, in that case, specific examples of letters help us to understand letters in general, so the example of weaving was introduced as an example of how to distinguish arts of division from arts of combining, all in the name—or with the goal, it now seems—of helping one to become better at the arts of dividing and combining more generally. In other words, the method will have been the message.[13] The discussion of weaving was undertaken not simply for its own sake but so as to make those who follow the discussion better able to think about other subjects. Weaving was used as an example to help them find the statesman, but also as an example of how to use examples in order to become better dialecticians.

But there are, of course, not only similarities but important differences between weaving and dialectics. Perhaps the biggest difference is that the former is a "material" or bodily practice while the latter is an immaterial one. As the Stranger argues, while some things have "sensible resemblances [*homoiotētes*] that are easy to perceive" and to point out, the "greatest and noblest conceptions" have no image [*eidōlon*] for human vision. Hence "we must [*dei*]," says the Stranger, in yet another plea for method endeavor by practice to acquire the power of giving and understanding a rational definition [*logon*] for each one of them; for immaterial [*ta asōmata*] things, which are the noblest and greatest, can be exhibited by reason alone [*logōi monon*], and it is for their sake [*heneka*] that all we are saying is said. But it is always easier to practice in small matters than in greater ones. (286a–b)

Hence one practices on weaving in order to arrive at even more noble things—"immortal things," which can be exhibited "by reason alone" (*logōi monon*) (286a–b), things such as statesmanship, perhaps, which will have at least a share of the immortal. Material examples are thus given to help us understand immaterial things, just as political regimes modeled on some bygone golden age can be used, perhaps, to help us understand the nature of life in that bygone age.

All this must be kept in mind, says the Stranger, when they ask whether their prior discourses were too long. They must not become impatient with all the talk about weaving and the revolutions of the universe or, as in the *Sophist*, about the existence of non-being. They must not simply look at the length of their discourses relative to other discourses and consider them to be too long or "irrelevant" (*perierga*)—that is, as ancillary or beside the point (see *Phaedrus* 274a). They must instead ask whether their discourses, though long, have been of a proper or appropriate length—that is, whether they have been proper or fit for their purposes. The Stranger can now affirm that the first criterion, in fact, for determining whether a speech's length is appropriate is whether it makes one a better dialectician or not. The end is or should not be either pleasure or efficiency but *method itself*, making one better at the dialectical method. One can ignore—except as a secondary consideration—whether a speech leads to a quick answer or whether it brings pleasure, or whether it persuades an audience, and ask simply whether it helps one become a better or worse practitioner of the method itself (286e). As the Stranger says, defending himself here against the criticism—a criticism that he himself voiced in order to answer—that they have been "beating around the bush," "going round in circles," or moving in a "roundabout" way.

By far our first and most important object should be to exalt the method [*methodon*] itself of ability to divide by classes, and therefore, if a discourse, even though it be very long, makes the hearer better able to discover the truth, we should accept it eagerly and should not be offended by its length, or if it is short, we should judge it in the same way. And, moreover, anyone who finds fault with the length of discourses in our discussions, or objects to roundabout methods [*en kyklōi periodous*], must not merely find fault with the speeches for their length . . . but must also show that there is ground for the belief that if they had been briefer they would have made their hearers better dialecticians and quicker to discover through reason the truth of realities [*tōn ontōn*]. (286e)

The way to determine the proper length of a discourse is thus to ask whether it promotes the method itself, for it is only by that method that the truth can be discovered. Truth, then, the possibility of approaching or arriving at the truth, is that for the sake of which a discourse is drawn out. Truth is or should be the measure of the proper length of a discourse, not pleasure or utility or, as we are about to see, *persuasion* or *efficacy*, even if these may be secondary considerations or ancillary benefits. These references to persuasion and efficacy already suggest that the *Statesman* is staging a contest, almost always just below the surface, between philosophy and sophistry (or rhetoric) over the values of force and life. If Plato came to be as critical of the sophists and rhetoricians as he was of the Heracliteans, it was, first and foremost, because none of these had an adequate conception of life—that is, a conception of life in conjunction with being and truth.

CHAPTER 5

Fruits of the Poisonous Tree: Plato and Alcidamas on the Evils of Writing

> Writing in the common sense is the dead letter; it is the carrier of
> death. It exhausts life. On the other hand, on the face of the same
> proposition, writing in the metaphoric sense, natural, divine, and
> living writing, is venerated; it is equal in dignity to the origin of
> value, to the voice of conscience as divine law . . .
>
> —JACQUES DERRIDA, *Of Grammatology*, 17

As I recalled in the introduction to this work, Jacques Derrida's seminal
1968 essay "Plato's Pharmacy" focuses on Socrates's critique of writing near
the end of the *Phaedrus* and the characterization of writing as a *pharmakon*. But "Plato's Pharmacy" is perhaps above all else an essay about the question of *life* in Plato's dialogues, about what gives life, what is on the side of
life or purports to be on the side of life, and what is or is thought to be on
the side of death. As Derrida shows, the theme or motif of *life* provides another way to think the internal coherence and structure of this seemingly
disjointed dialogue, another way to read its connection to other dialogues,
the theme of life, that is, in its relationship to breath, presence, power, paternity, fecundity, memory, dialectics, and so on—in other words, to the
entire matrix we call Platonism. In this chapter, I would like to read "Plato's Pharmacy" on this theme or question of life in order then to supplement the analyses we find there with a closer look at a figure and a text
Derrida mentions in passing on a couple of occasions, namely, Alcidamas
and his discourse *On the Sophists*. I will do this in order to contrast Plato's

critique of writing in the *Phaedrus* with Alcidamas's but, also, in order to bring us back to the *Statesman*. For the *Statesman* is perhaps the closest of all of Plato's dialogues to the *Phaedrus* and so deserves to be read along-side it, not only because it too combines in a single dialogue at once *diairesis*, myth, and reflections on method, to name just a few of their many common elements, but because it too revolves around the question of life, around the inferiority of written law to living discourse and the necessity of writing for life in the polis or a life in philosophy.

"Plato's Pharmacy" and Life

The question of life is at the center of "Plato's Pharmacy" from its very first page onward—that is, from the very brief, untitled, introductory section in which Derrida opposes the text or the work as organism, as living being, to the text as "woven texture" or as "web."[1] Such references to the text as web or textile could easily lead the reader to assume that Derrida's essay will focus not on the *Phaedrus* but on the *Statesman*, since weaving is there introduced, as we just saw, not only as a paradigm for statesmanship but, in conjunction with the example of writing, as a paradigm for paradigms. Just a couple of pages into "Plato's Pharmacy," Derrida writes:

> The example we shall propose of this will not, seeing that we are dealing with Plato, be the *Statesman*, which will have come to mind first, no doubt because of the paradigm of the weaver, and especially because of the paradigm of the paradigm, the example of the example—writing—which immediately precedes it. We will come back to that only after a long detour. We will take off here from the *Phaedrus*. We are speaking of the *Phaedrus* that was obliged to wait almost twenty-five centuries before anyone gave up the idea that it was a badly composed dialogue. ("PP" 65–66/74)[2]

Derrida's main reason for focusing on the *Phaedrus* rather than the *Statesman* is, of course, the former's emphasis on the opposition between speech and writing, particularly in the myth told by Socrates near the end of the dialogue. This opposition, on Derrida's reading, will reveal "a more secret organization of themes, of names, of words," a "whole *symplokē*" of arguments and terms, all of which revolve, as we will see, around the theme of life ("PP" 67).

The entire *Phaedrus* is motivated by Phaedrus luring Socrates out of Athens by means of a written discourse, a text hidden beneath his cloak, one that he, Phaedrus, had hoped to memorize and then try out on Socrates. Socrates will have none of this, of course, and so will in turn draw Phaedrus out by getting him to reveal the text, the discourse of Lysias, and then engage in a live conversation about it, a live conversation—a dialectical conversation—that will be about the differences between philosophy or dialectics and rhetoric, speech and writing, and, concomitantly, life and death. This emphasis on life becomes most explicit when, around three quarters of the way through the dialogue, just at the point he is comparing the composition of Lysias's discourse on love to his own, Socrates suggests, in a line we have already looked at, that "every discourse [*logon*] must be organized, like a living being [*zōon*], with a body of its own, as it were, so as not to be headless or footless, but to have a middle and members, composed in fitting relation to each other and to the whole" (*Phaedrus* 264c). A *logos* is, or should be, organized like a *zōon*, says Socrates, with a head and feet, parts or limbs, extremities and joints, that fit together into a whole— that is, into an *organic* whole. It's a metaphor, an analogy, but one that shapes and determines the entirety of Plato's dialogue—indeed, as Derrida will go on to show, the entirety of Plato's work, including the way in which metaphors or analogies are used and understood in the dialogues.

Now if, as Socrates argues, a *logos* is or should be organized as a *zōon* (that is, as an animate creature) then it must be engendered, have parents, or, at the very least, a single parent—a father. "Like any person," Derrida writes, "the *logos-zōon* has a father" ("PP" 80). Derrida will thus go on to demonstrate "the permanence of a Platonic schema that assigns the origin and power of speech, precisely of *logos*, to the paternal position" ("PP" 76). He will follow this figure of the father throughout Plato's dialogues, from Book 6 of the *Republic* where the Good is characterized as a Father and the Sun as his offspring to the myth of writing at the end of the *Phaedrus*, where Thamus, the king, situated in the place of the father, ultimately pronounces his judgment on writing, the invention of Theuth, who plays the role of the son and servant (see "PP" 82). Thamus's judgment is, of course, a negative one—a negative pronouncement spoken by the king-father in support of the spoken word. As Derrida will go on to show, the father in Plato is the source or origin not just of speech but of value more generally, including and especially the value of life—the *father*, notice, before or without or to the exclusion of the mother, which, of course, orients and determines everything.

It is the value of life, then, first exemplified in the figure of the father, that marks the opposition between speech and writing. As Derrida argues, "living *logos* is alive in that it has a living father . . . a father that is *present*, *standing* near it, behind it, within it, sustaining it with his rectitude, attending to it in person in his own name" ("PP" 77). The father not only gives life to his offspring but continues to give that offspring value and legitimacy through his living presence. The "specificity of writing," then, as opposed to speech or living logos, is "intimately bound to the absence of the father" ("PP" 77), to the absence of a *living* father who might defend his writing—that is, his offspring—when it is questioned or comes under attack. In contrast to living logos, writing is essentially fatherless, orphaned, without any living or life-giving force behind it ("PP" 145). As opposed to legitimate speech, it "substitutes the breathless sign for the living voice, claims to do without the father (who is both living and life-giving) of *logos*, and can no more answer for itself than a sculpture or inanimate painting can ("PP" 91–92). "It goes without saying," Derrida will conclude, "that the god of writing," in this case Theuth (the Egyptian equivalent of Hermes), "must also be the god of death" ("PP" 91).

Writing is thus presented by Socrates in the *Phaedrus* as lacking a living father to animate and defend it, a father who can give it value and legitimacy. But that is really just the beginning of Plato's critique of writing, which is called a *pharmakon* in the dialogue, a word that means either remedy or poison, depending on the context, or, as Derrida will demonstrate, at once remedy and poison.[3] As such, "the *pharmakon* and writing are always involved in questions of life and death" ("PP" 105). As a *pharmakon*, writing is harmful, according to Socrates in the *Phaedrus*, because it is artificial, because it "goes against natural life" and so is "the enemy of the living in general" ("PP" 100; see 104). Whereas living speech is able to sow seeds that can grow in the soul of another, writing, says Socrates, does nothing more than repeat the dead letter, producing written traces that do not belong to the order of *physis* and so must be artificially, mechanically repeated ("PP" 105). Speech and writing are thus opposed like two forms of memory, one bad, one good:

> While the phonic signifier would remain in animate proximity, in the living presence of *mnēmē* or *psuchē*, the graphic signifier, which reproduces it or imitates it, goes one degree further away, falls outside of life, entrains life out of itself and puts it to sleep in the type of its double . . . Instead of quickening life in the original, "in person," the *pharmakon* can at best only restore its monuments. ("PP" 110)

Rather than reanimating memory, rather than leading one back to the truth, as speech can do, the *pharmakon* of writing hypnotizes, fascinates, and memorializes. It leads to nontruth, nonknowledge, and death ("PP" 105). This is philosophy's indictment of sophistry or the sophists, who, knowing nothing of genuine, reanimating memory, have only monuments, inventories, recipes, readymade phrases, "formulas learned by heart," at their disposal ("PP" 73, 106). Living speech is therefore opposed by Plato not only to writing in the literal or restricted sense of the term but—and this will become crucial when we turn to Alcidamas in just a moment—to all kinds of learning by heart as they are practiced in what are commonly called *spoken* discourses.

The categories of life and death thus mark not only the difference between speech and writing but the difference between two types of memory and two types of repetition, one spontaneous and living, the reproduction in the present of a living truth, the other mechanical and mortifying, the mere repetition of a sign distinct from the truth: "Writing would indeed be the signifier's capacity to repeat itself by itself, mechanically, without a living soul to sustain or attend it in its repetition, that is to say, without truth's *presenting itself* anywhere" ("PP" 111). While writing is thus identified in Plato with a mere repetition of the signifier, speech is identified with a living memory that, as dialectics, through what Derrida calls the "living tension of dialectics" ("PP" 111), attempts to revive or reanimate the signified. There is, then, on the side of life, "'good' repetition (which presents and gathers being within living memory)" and, on the side of death, bad, mechanical repetition ("PP" 135).

At issue here is nothing less than truth itself in its relation to life—that is, on the one hand, "a repetition of truth (*alētheia*) which presents and exposes the *eidos*" and, on the other, "a repetition of death and oblivion (*lēthē*) which . . . re-presents a presentation, repeats a repetition" ("PP" 135). As such, writing is always identified with play and frivolity and should not be taken seriously; indeed it should be pursued only as an amusement, diversion, or pastime ("PP" 156–157). Much later in the essay, Derrida will characterize these two forms of repetition as two different ways of thinking tautology, one on the side of death and one on the side of life. On the one hand, "tautology is life only going out of itself to come home to itself. Keeping close to itself through *mnēmē*, *logos*, and *phōnē*; on the other hand, repetition is the very movement of non-truth . . . tautology is life going out of itself beyond return . . . [an] irreducible excess, through the play of the supplement, of any self-intimacy of the living, the good, the true" ("PP" 168–169). To put Derrida's entire argument in a single, untranslatable phrase, *"Le logos s'aime lui-même"* ("PP" 170).[4]

Writing lacks the living presence of a father; "it has no power [*dunatos*] to protect or help itself" and it is unable to engender living memory in another (*Phaedrus* 275d–e). Like painting, however, writing is able to give the *illusion* of life. As Socrates says to Phaedrus: "Writing has this strange quality [that is, this *deinon* quality], and is very much like painting, for the creatures of painting stand like living beings, but if one asks them a question, they preserve a solemn silence. And so it is with written words" (275d; cited by Derrida at "PP" 136). Both painting and writing are "impotent to represent a live word properly, to act as its interpreter or spokesman." They are "mere figurines, simulacra" ("PP" 136), simulacra of life that nonetheless give the appearance or even the illusion of life. So good are they at giving the appearance of life that Plato suspects they are not completely dead—yet another reason they are to be distrusted.[5] Writing may be sterile, impotent, unable to produce a living memory in another, but it nonetheless appears to have a certain force, a certain power of illusion, one that must be taken into account and guarded against. The *Phaedrus* is, it seems, an attempt to do just that; it is at once a warning, a séance, and an exorcism.

Hence the distinction between speech and writing, a distinction that could easily appear epiphenomenal in Plato's work, brings along with it, on Derrida's analysis, everything that is central to the dialogues: presence, memory, knowledge, legitimacy, truth, as well as their opposites, and all of these in relationship to life and death. Each of these would "form a system," as Derrida argues, "with all the great structural oppositions of Platonism" ("PP" 111), a system that exceeds the explicit themes or arguments of the dialogues, a system that links the values of life and death to all the other oppositions in the Platonic matrix. Derrida writes:

> Our intention here has only been to sow the idea that the spontaneity, freedom, and fantasy attributed to Plato in his legend of Theuth were actually supervised and limited by rigorous necessities. . . . Plato had to make his tale conform to structural laws. The most general of these, those that govern and articulate the oppositions speech/ writing, life/death, father/son, master/servant, first/second, legitimate son/orphan-bastard, soul/body, inside/outside, good/evil, seriousness/ play, day/night, sun/moon, etc. ("PP" 85)

Speech is thus opposed to writing as life is opposed to death or, in accordance with the "configurative unity of these significations" ("PP" 87), as the father is opposed to the son, the legitimate son to the orphan-bastard, the master to the servant, the soul to the body, inside to outside, good to evil, seriousness to playfulness, day to night, sun to moon, and so on. What we begin to glimpse in the myth of writing at the end of the *Phaedrus* is, on Derrida's reading, nothing less than Platonism itself.

But then what is to be made of the fact that, in the *Phaedrus*, living speech, the speech that is most alive, the speech spoken not with the voice but in the soul itself, the speech that is characterized at *Sophist* 263e as "a silent inner conversation [*dialogos aneu phōnēs*] of the soul with itself," is itself a kind of *writing*, namely, "writing in the soul"? As Phaedrus quickly sees, this writing is the "living [*zōnta*] and breathing [*empsychon*] word of him who knows [*tou eidotos*], of which the written word [*ho gegrammenos*]"— that is, what is commonly *called* the written word—"may justly be called the image [*eidōlon*]" (*Phaedrus* 276b).[6] Whereas Socrates had earlier argued that writing should never be pursued seriously but only as a sort of pastime, that it should simply supplement speech and "serve only to remind [*hypomnēsin*] us of what we know" (276d), another kind of writing—"writing in the soul"—now names the very thing of which writing as it is commonly understood should remind us. Derrida writes, expressing his astonishment and yet recognizing that this too is part of the program and the structure of Plato's dialogues:

> It is not any less remarkable here that the so-called living discourse should suddenly be described by a "metaphor" borrowed from the order of the very thing one is trying to exclude from it, the order of its simulacrum. Yet this borrowing is rendered necessary by that which structurally links the intelligible to its repetition in the copy, and the language describing dialectics cannot fail to call upon it. ("PP" 149)

It is by means of a metaphor of writing, then, that Plato opposes not only speech to writing but, now, a good writing to a bad one. And, once again, the distinction is made along the lines of life.[7] As Derrida writes, at once summarizing his reading of Plato and revealing some of the larger ambitions of that reading:

> According to a pattern that will dominate all of Western philosophy, good writing (natural, living, knowledgeable, intelligible, internal, speaking) is opposed to bad writing (a moribund, ignorant,

external, mute artifice for the senses). And the good one can be designated only through the metaphor of the bad one. ("PP" 149)

For Plato, dialectics is the antidote to writing, live speech the remedy to written signs. And yet Plato himself identifies this antidote as itself a kind of writing—that is, as "writing in the soul." Derrida will thus spend several pages analyzing Plato's frequent use of the example of letters (for example, the example of letters we just looked in the *Statesman*) and of the metaphor of writing (see "PP" 159–162). Writing, phonetic writing, would appear to be the necessary supplement to the supplement that voice or spoken discourse already is ("PP" 109), at once preserving and corrupting it ("PP" 137, 139). It is this reliance upon writing to signify what supposedly comes before writing, this turn to a metaphor of technology to designate what is thought to be most natural, that brings Derrida, at the end of "Plato's Pharmacy," though only in passing, back to the *Statesman* and to the "paradigm" of the *symplokē* ("PP" 165). It is here that we would have to return to this dialogue in order to think, with Derrida's essay as our guide, the relationship between writing and law, written law as the necessary supplement to spoken law or, better, to science, writing not just as second best, as what must substitute for a living lawmaker, but as necessary to the very notion of law itself. It is here that we would also, perhaps, have to think, with Derrida's help, the Age of Zeus and the imitation that is central to it as the necessary supplement to the Age of Kronos, the reign of the son as the necessary supplement to the reign of the father.

Alcidamas and the Superiority of Extemporaneous Speech over Writing

Another way, however, of returning to the themes of life and death, presence and absence, in the *Statesman*, would be to look at another critique of writing besides the one found in the *Phaedrus*, a critique of writing that Plato either influenced or was influenced by, a critique that, as we will see, can shed light not just on the *Statesman* and the *Phaedrus* individually but on the relationship between them. Early on in "Plato's Pharmacy," Derrida recalls this other critique of writing and points out that what it has in common with Plato's is nothing other than this emphasis on life that I have been tracking through Derrida's essay:

In describing *logos* as a *zōon*, Plato is following certain rhetors and sophists before him who, as a contrast to the cadaverous rigidity of writing, had held up the living spoken word, which infallibly conforms to the necessities of the situation at hand, to the expectations and demands of the interlocutors present, and which sniffs out the spots where it ought to produce itself, feigning to bend and adapt at the moment it is actually achieving maximum persuasiveness and control. ("PP" 79)

As Derrida makes clear in a footnote, the two orators he has in mind are Isocrates and Alcidamas, two orator-sophists whom Derrida mentions, often in the same breath, without actually citing, at several key junctures of his essay (see "PP" 79n12, 112, 113–114; see also 148–149 for a mention of Alcidamas alone).[8] A closer look at the critique of writing in these two thinkers, and particularly in Alcidamas, will demonstrate that while both share with Plato a similar concern for the values of life and the living, they have, in the end, a very different conception of what life is and what values it favors. In other words, what will distinguish Plato from the sophists, or at least from *these* sophists, will be a difference over the value of rhetoric, to be sure, but also, and more importantly, a difference over the value of life.

At the beginning of his discourse *On the Sophists*, which also bears the title "on those who write written discourses," Alcidamas makes it clear that his criticism is aimed not at sophistry as a whole but at those sophists who have devoted their lives to mastering writing at the expense of speech.[9] These men, he says, are "woefully deficient in rhetoric and philosophy" and so should probably be called "poets rather than Sophists," by which he means *writers* of artificially constructed poetic phrases rather than *speakers* of extemporaneous prose (*OS* 2). Though writing is not to be condemned altogether, it should, he says, echoing Socrates at *Phaedrus* 274a, be "practiced as an ancillary pursuit [*parergōi*]" (*OS* 2).[10]

What follows in *On the Sophists* is a long list of accusations against writing, some of which can be found in Plato but some not. For example, Alcidamas underscores even more than Plato, though the point is hardly absent from the *Phaedrus*, the facility of writing as opposed to extemporaneous speaking. "Writing is easier than speaking" (*OS* 5), he says, because the writer has the leisure to meditate upon his words and arguments and is able to rewrite what he has written. While few men are thus able to master extemporaneous speaking, many—just about anyone, it seems—can learn how to write. "Easily and readily practiced by anyone of ordinary ability," writing is essentially democratic ("PP" 144), which makes it an object of suspicion if not scorn for both Plato and Alcidamas. As Alcidamas says, and Plato would have no doubt agreed, "things good and fair are ever rare and difficult to acquire" (*OS* 5). While a good extemporaneous speaker could, with just a bit of effort, become a good speechwriter, the opposite is not the case (*OS* 6). In fact, says Alcidamas, and this claim very much echoes Plato's characterization of writing as a *pharmakon*—that is, as a *poison* for living memory—spending too much time writing speeches can actually make one less able to speak extemporaneously, more prone, says Alcidamas, to "suffer mental embarrassments, wanderings, and confusion" (*OS* 8).[11] The best proof that extemporaneous speech is to be considered more valuable than speechwriting is the fact that the best speechwriters actually attempt to "imitate the style of extempore speakers" and make speeches that "least resemble written discourses" (*OS* 13).

Speaking extemporaneously on a given subject is thus harder than writing a speech on that subject, though, interestingly, it is easier to deliver an extemporaneous speech than a written one, for "to memorize topics is easy, but to learn by heart an entire speech, word by word, is difficult and onerous" (*OS* 19). Writing speeches is both easier and harder; it is easier because one has the leisure to write and rewrite, and yet harder because, once written, a speech must be memorized down to the letter and any forgetting of words or of the order of arguments will lead to perplexity and embarrassment.

Like Plato, Alcidamas criticizes written speeches for their stiltedness, their artificiality, their lack of truth and spontaneity. But, unlike Plato, Alcidamas is concerned first and foremost with the effect this lack of truth will have on an audience. "The truth is that speeches which have been laboriously worked out with elaborate diction (compositions more akin to poetry than prose) are deficient in spontaneity [*automaton*] and truth [*alētheiais*], and, because they give the impression of a mechanical artificiality and labored insincerity, they inspire an audience with distrust and ill-will" (*OS* 12). Like Plato, then, Alcidamas condemns the written word as a mere simulacrum of living speech, a semblance or imitation of it. But Alcidamas focuses much more than Plato on writing's lack of efficacy and practicality. As Alcidamas says in a passage that may have inspired the *Phaedrus*, unless it's the other way around:

> Written discourses, in my opinion, certainly ought not to be called
> real speeches, but they are as wraiths [*eidōla*], semblances [*schēmata*],
> and imitations [*mimēmata*]. It would be reasonable for us to think of
> them as we do of bronze statues, and images of stone, and pictures of
> living beings [*zōon*]; just as these last mentioned are but the semblances
> of corporeal bodies, giving pleasure to the eye alone, and are of no
> practical value. (*OS* 27)

Alcidamas appears to criticize writing, just as Plato does, for its lack of reality, but his concern is much less to deliver an ontological or epistemological critique of writing, to question its legitimacy, its pedigree or its relation
to truth, than to question its lack of *practicality* or *force*. Alcidamas writes,
linking the value of life not to truth and reality, as Plato did, but to efficacy, to force or *energeia*.

> Just as the living human body has far less comeliness than a beautiful
> statue, yet manifold practical service, so also the speech which comes
> directly from the mind, on the spur of the moment, is full of life [*zēi*]
> and action, and keeps pace with events like a real person, while the
> written discourse, a mere semblance of the living speech, is devoid of
> all efficacy [*energeias*]. (*OS* 28)

This emphasis on force, on liveliness, is, of course, not so surprising for a
disciple of Gorgias, for whom logos in the form of persuasive eloquence is
a kind of witchcraft, a more powerful *pharmakon* than writing. As Derrida
writes in "Plato's Pharmacy," drawing Gorgias, Isocrates, and Alcidamas
all together into the same circle:

> the Attic School (Gorgias, Isocrates, Alcidamas) extolled the force of
> living *logos*, the great master, the great power: *logos dunastēs megas estin*,
> says Gorgias in his *Encomium of Helen*. The dynasty of speech may be
> just as violent as that of writing, but its infiltration is more profound,
> more penetrating, more diverse, more assured. The only ones who
> take refuge in writing are those who are no better speakers than the
> man in the street. Alcidamas recalls this in his treatise . . . "on the
> Sophists." ("PP" 115)[12]

According to Alcidamas, extemporaneous speech is more true than writing, but also, and more importantly, more useful, more powerful, and
more effective. As in Plato, the difference between speech and writing is
framed in terms of life and vitality, but the notions of life and vitality at

issue here are very different from the ones we find in Plato. For Alcidamas, it is the speaker's ability to respond in the moment, to keep pace with the changing circumstances of life, to intervene in life through persuasive speech, that marks the real specificity and superiority of extemporaneous speech. Derrida introduces this aspect of the difference between speech and writing in this way:

> For Isocrates, for Alcidamas, *logos* was also a living thing (*zōon*) whose vigor, richness, agility, and flexibility were limited and constrained by the cadaverous rigidity of the written sign. The type does not adapt to the changing givens of the present situation, to what is unique and irreplaceable about it each time, with all the subtlety required. While *presence* is the general form of what is, the *present*, for its part, is always different. ("PP" 113–114)

For Isocrates and, especially, for Alcidamas, speech, presence, and life are all aligned, just as they are in Plato. But because this entire configuration is oriented not toward the repetition of truth in the soul but toward the production of *persuasion*—that is, toward swaying an audience in the present moment, toward force and efficacy—everything takes on a new value. As Derrida writes, still lending his voice to this rhetorical or sophistical critique of writing:

> writing, in that it repeats itself and remains identical in the type, cannot flex itself in all senses, cannot bend with all the differences among presents, with all the variable, fluid, furtive necessities of psychagogy. . . . In attending his signs in their operation, he who acts by vocal means penetrates more easily into the soul of his disciple, producing effects that are always unique, leading the disciple, as though lodged within him, to the intended goal. It is thus not its pernicious violence but its breathless impotence that the sophists held against writing. ("PP" 114–115)

What the sophists held against writing, in short, is that it shows no signs of life and is of no value for intervening effectively within life. The most telling sign of extemporaneous speech's attachment to life and writing's detachment from it is the former's superior ability to respond appropriately to the moment, in short, to take advantage of the *kairos*. Alcidamas writes:

> To speak extemporaneously, and appropriately to the occasion, to be quick with arguments, and not to be at a loss for a word, to meet the situation successfully, and to fulfill the eager anticipation of the

audience and to say what is fitting [*tōi kairōi*] to be said, such ability is rare, and is the result of no ordinary training. (*OS* 3)

While good extemporaneous speakers are able to intervene as the occasion requires, writing "gives aid too late to save the day." Only extemporaneous speakers are able to gauge and respond to their audience in order to take advantage of the situation, to seize the moment and be effective, to hold sway and exercise force.[13] As a result, speaking is harder but more useful and efficacious than writing. Alcidamas concludes, "What sensible man, therefore, is envious of this ability to compose speeches—an ability which fails so completely at the critical moment [*tōn kairōn*]" (*OS* 10). Unable to respond as extemporaneous speech can to changing circumstances, unable to rewrite on the spur of the moment what has already been written, writing misses the *kairos* and cannot take advantage of the opportunities that fortune herself sometimes presents (*OS* 25). "What sensible person," Alcidamas concludes, "would approve of a practice which militates against the use of the help which fortune [*tuchēs*] gives?" (*OS* 26)

Extemporaneous speech, unlike writing, is able to take advantage of the opportune moment, to say what is fit or appropriate to the situation. Alcidamas and Isocrates both underscore this on multiple occasions.[14] Even Socrates in the *Phaedrus* seems to concede to rhetoric—that is, to what is commonly called rhetoric, an art that would include extemporaneous speaking—an ability to distinguish "the favorable occasions [*kairous*] for brief speech or pitiful speech or intensity and all the classes of speech which [the rhetorician] has learned" (*Phaedrus* 272a).[15]

But then there's this, which leads us back by other means or other terms to the *Statesman*: By being present to one's situation, by being able to take one's audience's reactions into account, extemporaneous speaking, says Alcidamas, allows one to avoid the extremes of excess or deficiency in one's speech:

> extemporaneous speakers exercise a greater sway over their hearers than those who deliver set speeches; for the latter, who have laboriously composed their discourses long before the occasion, often miss their opportunity [*kairōn*]. It happens that they either weary their listeners by speaking at too great length, or stop speaking while their audience is fain to hear more. (*OS* 22)

Because it is, as Alcidamas says, "difficult, if not impossible, for human foresight accurately to estimate the disposition of an audience as to the length of a speech," "the extemporaneous speaker has the advantage of being able to adapt his discourse to his audience; he can abbreviate or extend at will" (*OS* 23).[16]

Plato and Alcidamas on the Force of Speech and Writing

To speak appropriately, to take advantage of the critical moment, to lengthen or abbreviate one's speech based on the attention of the audience: These values of extemporaneous speech are not claimed for philosophy, as one might have expected, by Socrates's rehabilitation and appropriation of rhetoric in the *Phaedrus*, though they are reclaimed and reappropriated for statesmanship if not philosophy in the *Statesman*, as if this latter dialogue were the fitting, if not necessary, supplement to the *Phaedrus*.[17] In the *Statesman*, we recall, the Stranger introduces the distinction between two kinds of measures in order to determine whether their discourse, their logos, was too long, too short, or of the proper length. It is as if the entire distinction that is then drawn there is a response to the discourses of Isocrates and Alcidamas on the value of the measured, the opportune, the proper, and so on, as if the distinction were introduced in this other dialogue in order to fill out Socrates's passing comment in the *Phaedrus* that Prodicus had "discovered the art of proper speech, that discourse should be neither long nor short, but of reasonable length [*metriōn*]" (267b).[18] It is the question of the appropriate length of discourse—this time, significantly, a philosophical discourse—that compels the Stranger in the *Statesman* to distinguish, as we saw in the previous chapter, between two different forms of measurement, one concerned with "relative greatness and smallness" and the other with "something without which," says the Stranger, "production would not be possible." This latter is the measure of the mean, a form of measure that includes such categories as "the moderate [*to metrion*], the fitting [*to prepon*], the opportune [*ton kairon*], the needful [*to deon*], and all the other standards that are situated in the mean between the extremes" (284e)— that is, just those things that Isocrates and Alcidamas claim to be the unique privilege of extemporaneous speech. The explicit reason for taking this detour in the *Statesman* is, according to the Stranger, a concern over productivity in general, for no one would be able to do or produce anything, he argues, without this form of measure. But this conjunction of themes suggests that Plato is also in conversation in both *Statesman* and *Phaedrus* with Isocrates and, perhaps especially, with Alcidamas. If Plato will thus extend the use of terms such as the moderate, the proper, the opportune far beyond oratory or rhetoric, indeed into the arts more generally, into statesmanship, and perhaps even into philosophy as dialectic, oratory or rhetoric seems to have been their original provenance.[19]

It is thus more than a coincidence, I would argue, that the very terms the Stranger identifies with one form of measure are precisely those used by Isocrates and Alcidamas to condemn writing and extol the art of extemporaneous speechmaking. For Plato could not simply concede to sophistry such terms as the appropriate, the opportune, the moderate, and so on. These terms are so positively valued they just had to be appropriated by Plato for philosophy. Plato is able to do this, as is his wont, by detaching these terms from their original identification, with extemporaneous speech in order then to reattach them to dialectic (or at least to statesmanship) and thereby establish another discourse of life in opposition to Alcidamas's in particular and sophistry's more generally. The battle or gigantomachia between Plato and the sophists will then revolve around the relative values of speech and writing in relationship to memory, repetition, and, especially, life—all those things that Derrida tried to think together in "Plato's Pharmacy."

All this becomes even clearer when we look at the one term Plato will *never* attempt to appropriate for philosophy, the one term he will leave to the sophists after having identified it with everything that, in the *Phaedrus*, is attached to writing—that is, to the bad kind of writing. That is the term, paradoxical as it may initially seem, for extemporaneous speech itself, *autoschediazein*. Based on the verb *schediazein* (that is, to make or do a thing off-hand, to speak off-hand) the term *autoschediazein* is used no fewer than seventeen times in the nine pages of Alcidamas's *On the Sophists*, compared to just eight times in all of Plato's dialogues—a quantitative difference that will reveal a genuinely qualitative divide between the two thinkers on the question of speech and writing in relation to life.[20]

Apart from *Euthyphro* and *Apology*, which are generally considered to be early dialogues, where the word refers to acting or doing something in a careless, haphazard, or unadvised way, a way that is without precedent or preparation or forethought, *autoschediazein* is used in the dialogues exclusively in relation to speech and, arguably, exclusively in the register of oratory or rhetoric, where it means, just as we saw in Alcidamas, to speak extemporaneously, to innovate or to improvise in discourse. That is precisely how the term appears to be used in the *Phaedrus*. When Phaedrus encourages Socrates early on in the dialogue to try to produce a speech that would rival Lysias's, Socrates says with his characteristic self-deprecation, an ironic self-deprecation that, as we know, will be contradicted by everything that follows: "I shall make myself ridiculous if I, a mere amateur [*idiōtēs*], try without preparation [*autoschediazein*] to speak on the same subject in competition with a master of his art" (236d). Socrates appears to be saying a couple of different things here. By calling himself an amateur, an *idiōtēs*, he is suggesting that he is not a professional orator or speech-maker as Lysias is and so will be unable to compete with him. But he can also be heard to be contrasting extemporaneous speech, a speech that must be improvised and *spoken* on the spot, in the moment, with Lysias's carefully prepared *written* discourse, a discourse that Lysias would have been able to prepare at his leisure and so would have been able to re-arrange and rewrite at his leisure. Phaedrus, familiar with the language of rhetoric and oratory, would have readily recognized this term and all the values attached to it. Speaking *autoschediazein*, speaking extemporaneously, would be, as Alcidamas argues, harder than writing a speech. It thus seems as if Socrates is in agreement with Alcidamas regarding the relative value and difficulty of speechwriting in relation to extemporaneous speaking.

Elsewhere, however, Socrates—and, by implication, I think, Plato—will disabuse us of the impression that extemporaneous speech is to be positively valued by actually identifying it, paradoxical as this may seem at first glance, with writing—that is, with the bad kind of writing and the bad kind of memory and repetition to which it is attached in the *Phaedrus*. This happens in the *Menexenus*, as Socrates argues that speaking extemporaneously does not mean to speak "without preparation," completely spontaneously, but to speak in the moment by ordering words, phrases, and topoi that have already been prepared and memorized to a large extent beforehand.[21] Menexenus says to Socrates as they discuss those who, like Pericles in his famous funeral oration, will be called upon to deliver a speech to the Athenians: "You are always deriding the orators [*rhētoras*], Socrates. And truly I think that this time the selected speaker will not be too well prepared; for the selection is being made without warning, so that the speaker will probably be driven to improvise his speech [*autoschediazein*]." To which Socrates responds: "Why so, my good sir? Each one of these men has speeches readymade [*logoi pareskeuasmenoi*]; and what is more, it is in no wise difficult to improvise [*autoschediazein*] such things" (*Menexenus* 235c–

d)—especially when, as Socrates will go on to say, the speaker will be praising the Athenians in front of the Athenians.

Socrates thus suggests in the *Menexenus*, and demonstrates through his own speeches in the *Phaedrus*, that it is, in fact, not so difficult to speak *autoschediazein* because those who are accomplished in extemporaneous speech are never wholly unprepared but bring to every occasion for discourse a whole stock of time-tested tropes, topoi, recipes, and phrases that can all be deployed to sway and win the support of the audience. Socrates's description of rhetoric in the *Phaedrus* suggests that all these things were, in fact, commonly studied and committed to memory in order then to be used as the occasion warranted. That extemporaneous speaking is indeed very different from completely unprepared speaking is confirmed by Alcidamas himself when, near the end of *On the Sophists*, he imagines himself being accused of inconsistency by writing a discourse commending extemporaneous discourses over written ones and by "deeming chance to be of more worth than forethought, and careless speakers to possess greater wisdom than careful writers" (*OS* 29). To this objection, Alcidamas says he does not deny that "great preparation is required for a good and effective speech"; he is thus not at all "encouraging *careless* speaking" when he says that he "esteem[s] the ability to speak extemporaneously more highly than the written word" (*OS* 33). The accomplished extemporaneous speaker does indeed have a whole arsenal of phrases, topoi, and rhetorical strategies at his disposal. He has them memorized, at the ready; it's just that he has not memorized them in a particular, unchanging order.[22] Unlike the writer, or the person who has memorized a written speech, he remains open to the shifting moods of his audience, ready to use particular examples or phrases or tropes when the situation calls for them, ready to change the order of arguments, to shorten or lengthen his discourse, in order to take advantage of the occasion. But that does not mean that he is not relying upon a kind of mechanical, artificial memory—the very thing Socrates condemned under the name of *writing* in the *Phaedrus*. Both Alcidamas and Plato understand the extemporaneous speaker in just this way, and that is what makes him commendable in the eyes of the first and blameworthy in the eyes of the second. In short, speaking extemporaneously, speaking *autoschediazein*, is, for Plato, a kind of writing—that is, the bad kind, when compared to dialectic as writing in the soul. That is what allows Socrates in the *Menexenus*, this time without self-deprecation, to criticize those who speak *autoschediazein* in terms that are not wholly different from those used by Alcidamas to criticize speechwriters: it's easy to do what they do, Socrates can say about extemporaneous speakers; what's really hard and what has real value, what is most living, most rare, is what the philosopher or dialectician does.[23] Plato can thus share in the critique of writing as we find it in Isocrates or Alcidamas without having then to extol the virtues of extemporaneous speech. Both writing and extemporaneous speech are, in the end, fruits of the same poisonous tree and they both lead away from the source and essence of life itself.

Both Alcidamas and Plato condemn writing, but they do so by opposing it to two very different things: Alcidamas to extemporaneous speech and Plato to dialectics, which Plato then has the bravado to call not "extemporaneous speech in the soul" but, precisely, "writing in the soul." The difference in strategies is telling. Whereas Alcidamas, in his praise of extemporaneous speech, does not think written speeches really even deserve the name speeches, *logoi* (*OS* 27), Plato will go so far as to call true speeches, the most genuine and most valued ones, the kind that have their place in the soul, a certain kind of writing. Plato's strategy is not only dialogical but, it seems, dialectical, even if it also conforms, as Derrida argues, to a form of thinking that identifies writing as both the *mere* supplement and the *necessary* supplement to speech.[24] As Derrida remarks of Plato's critique of writing, "Alcidamas said more or less the same thing. But it marks a sort of reversal in the functioning of the argument. While presenting writing as a false brother—traitor, infidel, and simulacrum—Socrates is for the first time led to envision the brother of this brother, the legitimate one, as *another sort of writing*: not merely as a knowing, living, animate discourse, but as an *inscription* of truth in the soul" ("PP" 148–149).

Various kinds of spoken speech must thus be distinguished on the basis of their relationships to writing, on the one hand, and life, on the other. There is, first, the spoken speech that is a word-for-word repetition or recitation of a written one, a reading aloud or reciting by heart that makes the speaker sound like a book or like an automaton, the kind of speech Phaedrus proposes to try out on Socrates (a recitation of Lysias's speech) at the very beginning of the *Phaedrus*. At the other extreme, there would be, if it is possible to imagine this, an absolutely unprepared, careless form of speaking, a speaking for which Plato will have as little sympathy as Alcidamas. There is also, of course, inspired, enthused speech, something for which Plato might have a bit more sympathy and that the myth attempts to appropriate for philosophy, even though such speech must never be confused with philosophy. Then there is so-called extemporaneous speech, which is always somewhat prepared, always a kind of *hypomnēsis*, on Plato's view—that is, a kind of bad writing. Finally, there is dialectics or "dialectic method [*dialektikēi technēi*]," that most interior of spoken discourses, so interior and so immutable, so stable, that Plato will identify it as *writing* in the soul. This is, for Plato, despite, or perhaps because of, the "metaphor" of *writing*, the liveliest kind of speech, for it "plants and sows in a fitting soul intelligent words which are able to help themselves" and which are "not fruitless, but yield seed from which there spring up in other minds other words capable of continuing the process for ever [*aei athanaton*], and which make their possessor happy, to the furthest possible limit of human happiness" (277a). Hence Plato opposes this writing in the soul, this notion of an inner speech before writing, in short, dialectics, not only to speechwriting, as Alcidamas does, but to what Alcidamas calls extemporaneous speech, insofar as it, just like written speech, relies upon mechanical memory and repetition, upon formulas and recipes. In the absence of any genuine knowledge, in the absence of any animating, knowing presence, it may bring about persuasion in an audience through flattery and pandering but never any genuine knowledge. The antidote to the pharmakon of writing, the cure for this strange discourse that lacks a "father" and does not know how to respond when questioned, is thus not, for Plato, extemporaneous speaking, as it is for Isocrates and Alcidamas, but dialectics. By putting *both* writing and extemporaneous speech on the one side, and dialectic on the other, we have a better sense, I think, of what Derrida means when he says that "what Plato *dreams* of is a memory with no sign," that is, I think it could be said, a living memory without death ("PP" 109), a notion of life that is inseparable from knowledge and truth.[25]

But Plato has one more trick up his sleeve and one more term to appropriate. Socrates will claim that unless one *knows* the nature of the soul of the person one is trying to persuade, one will be doing nothing in accordance with art and, in the end, one's speech will not be truly effective. Like extemporaneous speech, dialectics will take advantage of the critical moment, but it will do so not in order to persuade—at least not first and foremost—but in order to *educate*. It is in this way that Socrates denies not only the truth of extemporaneous speech but its capacity for true *efficacy*—that is, its genuine liveliness. Real efficacy, real force, we are supposed to see, is exercised not by rhetoric but by dialectic. Rhetoric might supplement dialectic and so augment its force, but rhetoric without dialectic has no *real* force at all.[26]

What is at issue, then, in the confrontation between Plato and the sophists is nothing less than two different values of life. The *Phaedrus* and *Statesman* bear witness to this confrontation or this battle between two discourses of life—as well as memory, repetition, presence, and so on: the one identifying life with presence and power in the polis, the other with truth and philosophy. Two discourses of life: life as force and life as truth or, rather, since Plato would not want to give up or give up on any of these terms, including force, life as mere force and life as the force, the living force, of truth.

Alcidamas—In Memoriam

But what, then, is to be made of the uncanny force of writing, a force that Plato seemed at once to fear and to covet, which is no doubt why he dared to appropriate the term *writing* but not *autoschediazein* in order to characterize what is most precious for him, namely, thought as *writing in the soul*? And what are we to make of the fact that Alcidamas, too, seems to have recognized without recognizing—that is, without admitting to it—the power of writing, its extraordinary and somewhat uncanny power to extend speech in space and perpetuate it in time? While writing is associated throughout *On the Sophists* with a lack of force and efficacy, with a lack of life and an inability to respond to life, there is the tacit acknowledgement that writing perhaps has another kind of force or, perhaps, another kind of life. Up until the final paragraphs of *On the Sophists*, speechwriting appears to be a mere supplement, a by-product or a par-ergon, for those who know how to control it, a pastime that might, on occasion, supplement the serious business of speaking *autoschediazein*. As Alcidamas says— the Alcidamas who boasted of being able "to speak opportunely and felicitously on any subject proposed" (*OS* 31):

> Everyone knows that the ability to speak on the spur of the moment is necessary in harangues, in the courtroom, and in private conversation. It often happens that unexpected crises occur when those who can say nothing seem contemptible, while the speakers are seen to be honored by the listeners as possessors of god-like minds. (*OS* 9)

But in the final paragraphs of *On the Sophists*, Alcidamas begins to see other possibilities for written discourse. Unlike spoken discourses, which are easily forgotten and then hard to compare, a written discourse, he says, allows one to look into it at some later point, "as in a mirror," to "behold the advance of intelligence" (*OS* 32). Alcidamas sees in writing the chance for himself to look at himself in order to compare and to judge the relative value of his discourses, past and present. Writing, unlike speech, allows him to see different discourses at the same time and bring together discourses from different times—no small benefit of this seemingly worthless practice of writing.

But then Alcidamas, adding advantage to advantage, as it were, seems to recognize that writing also allows one to be honored not only in the present but in the future (which is perhaps to be honored as something even more than a god) and to be seen not only by oneself, as in a mirror, but by others as in a portrait. As Alcidamas writes, or confesses, and this would be his last reason for writing *On the Sophists*, "Finally, since I am desirous of leaving behind a memorial of myself, and am humoring my ambition, I am committing this speech to writing" (*OS* 32). Alcidamas says he has written the present discourse against writing because he wishes to leave something of himself behind, an acknowledgment, it seems, of writing's superior force over speech, writing's ability to live beyond life and address those to whom the writer was, precisely, never present, indeed writing's ability to exercise a force or power over those whom the writer never even knew, those who, by whatever chance or fortune, will have simply stumbled upon it—like a memorial or a funeral stele. Hence Alcidamas, perhaps like Plato, comes to recognizes the power of writing over speech in a speech that extols the virtues of extemporaneous speech over writing, a speech that, and there is more than just irony in this, is perhaps the only thing that today prevents the name Alcidamas from falling into oblivion.

As Derrida writes in "Plato's Pharmacy" of *tupoi*, that is, of written signs, "They will represent him even if he forgets them; they will transmit his word even if he is not there to animate them. Even if he is dead" ("PP" 104). Non-living signs that nonetheless transmit the words of a once living being: this is perhaps the only possibility that remains for those living in the Age of Zeus, the only force that remains for those living in a world from which they will one day be absent, that is, a world from which they are already, in principle if not yet in fact, separated by death.

CHAPTER 6

The Life of Law and the Law of Life

It is indeed on the side of chance, that is, the side of the incalculable
perhaps, and toward the incalculability of another thought of life, of
what is living in life, that I would like to venture here under
the old and yet still completely new and perhaps unthought
name "democracy."

—JACQUES DERRIDA, *Rogues*

Statesmanship *and* Epistēmē

Having demonstrated that we learn by example, and having then proposed
weaving as an appropriate example for understanding the art of statesmanship, the Stranger now applies this example to statesmanship in order to
bring his and the Young Socrates's search to a conclusion (287b–293e). They
have already separated off statesmanship or the art of the king from its kindred arts (those having to do with the herding of other kinds of animals), just
as they separated off weaving from its kindred arts (the making of rugs,
shoes, and so on). Now they must separate off the art of the king from the
co-operative arts, those that are either contingent causes or actual causes,
just as they separated off weaving from its contingent causes, the arts that
make tools for the loom, and its actual causes—for example, twisting thread,
which is a cause of making clothes or weaving but not weaving itself.

To distinguish, then, the contingent causes from the actual causes, the
Stranger suggests—by means of an analogy that we looked at in some de-

tail in Chapter 1—that they divide the contingent causes not in two, this time, but "divide them like an animal that is sacrificed, by joints" into a number of parts as close to two as possible (287c). They thus come up with seven categories or classes of possessions in the state, and thereby seven contingent causes of statesmanship, seven classes of things that are essential for the existence of the state and for statesmanship but that must not be confused with statesmanship itself: 1. tools (for production); 2. receptacles (for the preservation of what is produced); 3. vehicles (for transport); 4. defenses (clothes and arms for the individual, walls and buildings for the state); 5. playthings (music, painting, ornamentation, imitations, all those things done or produced for pleasure, for "play" rather than for "serious purpose" [288c]—including, we would have to conclude on the basis of this description, all *writing*); 6. natural resources (wood, metals, barks, skins)—a category of contingent cause that, says the Stranger, could have been cited first because it provides the raw materials for all the other contingent causes; 7. nourishment (all those things, such as farming, hunting, gymnastics, medicine, and cooking, that aim at the health of the body).

This list of seven classes contains all the property or possessions of the state except, says the Stranger, tame animals (289a). Everything but tame animals can be included—or forced, shoehorned, we might say— into one of these seven classes.[1] Left, then, only with "tame animals" (a distinction, we recall, that was assumed in a previous *diairesis* and had to be drawn out explicitly later [see 263e–264a]), the task now becomes separating off those within this class who will contest the art of the statesman. That is, with the contingent causes removed, they must now separate off those real or actual causes that are not part of statesmanship proper.

They will begin, says the Stranger, by dividing up those "tame animals," namely, human beings, that were not among the animals already listed among the possessions of the state.[2] The Stranger says he "prophesies" that amongst these they will find some who will contest the king for his "fabric" (*plegma*)—just as the "carders" and "spinners" did earlier with weaving. They begin with all those who make no claim to statesmanship or the kingly art, "bought servants," that is, "slaves," who do not serve the state voluntarily, and then, among "free men," those who do serve the state voluntarily as merchants, shipmasters, peddlers, and so on, all those who sell products at home or abroad. None of these lay claim to the art of statesmanship, and the same goes for heralds, clerks, and others who can rightly be called "servants" but not at all practitioners of statesmanship.

The Stranger feigns at this point not to know who these pretenders to statesmanship might be and he wonders aloud whether it was not just a "dream" that made them think they would ever find them. But he then says that he can now see a category of servants who, out of "pride and esteem," may indeed be these pretenders to statecraft. These are the class of diviners who act as the "interpreters" (*hermeneutai*) of gods to men, prophets and priests who know how to offer sacrifice and prayer (290c). Among both Egyptians and Greeks, says the Stranger, such men are held in great esteem; indeed in Egypt only a priest can rule, and, in Athens, "the holiest and most national of the ancient sacrifices are performed by the man whom the lot has chosen to be the King" (290e)—in other words, the King Archon, the one to whom Socrates went to answer the indictment on the previous day.

It is among this class of priests and prophets and king archons that the pretenders to statesmanship might be found, in addition, the Stranger says, to another class that has just "come in sight," the sophists, a "queer" lot, a "mixed race," some looking like lions or centaurs, fierce beasts, and some like satyrs and "cunning beasts" (*polutropois thēriois*), men who "make quick exchanges of forms and qualities [*ideas . . . dunamin*] with one another" (291a–b). The kind of sophist who concerns himself with the affairs of the state is, of all these, "the greatest charlatan [*goēta*]" (291c), in short, the Stranger is suggesting, the greatest shaman or sham. It is essential, then, that he, before all the others, be separated off from the real statesman and king.

But instead of going on to do just this, instead of distinguishing the statesman from all these pretenders to statesmanship (interpreters, priests, prophets, and, especially, sophists), the dialogue takes yet another abrupt turn. The Stranger says they must now distinguish (anticipating, in this regard, Aristotle's *Politics*) six forms (*schēmata*) of political rule or government, depending on whether there is *one*, *few*, or *many* rulers, and on whether the rule is voluntary (based on "voluntary obedience" and law) or compulsory (based on "enforced subjection" without law). Divided up in this way, we end up with six different possible regimes or constitutions— three where rule is over voluntary subjects with law: royalty or monarchy (rule by one), aristocracy (rule by few), and democracy (rule by many or the multitude), and three where rule is over unwilling subjects without law: tyranny (rule by one), oligarchy (rule by few), and democracy (rule by many—with the many being defined as a mob rather than a multitude).[3]

Which of these six forms of rule, then, is the "right" (*orthēn*) one? One, few or many, with or without force, with or without laws and a written constitution? Rather than answer this question, the Stranger makes yet another move that takes us back to that very first *diairesis* at the beginning of the dialogue. He recalls that the royal art was earlier called a science—an *epistēmē*—namely, a science of "judgment and command" that rules over living beings (292b). This was, recall, more or less stipulated at the very outset of the dialogue and everything up to and including this moment in the dialogue has depended on or assumed this. But the Stranger now remarks that they have not yet determined this science with sufficient accuracy. It is with regard to this science and not the number of rulers (the one, the few, or the many) or the mode of rule (voluntary obedience or enforced subjection, rule with or without law) that the distinction between various forms of government ought to be made. In other words, science and science alone should be what distinguishes these various forms of government; only science should determine who the true statesman is. That question then becomes: in which of these "forms of government . . . is engendered the science [*epistēmē*] of ruling men, which is about the greatest of sciences and the most difficult to acquire"? (292d). Without a knowledge of this science, without this *epistēmē*, they will never be able to distinguish the true statesman—that is, the one who truly *is* a statesman—from those who merely pretend "and make many believe [*peithousi*]," speaking persuasively and, considering what we saw in the last chapter, taking advantage of the *kairos*.

The Stranger now engages in some reasoned speculation about this science (292e). Because this science is so difficult, it is impossible for a "multitude" to acquire it. Appealing to common experience, he notes that few people ever become really good at any one thing. Hence a city of a thousand could hardly produce fifty great "draught-players," let alone fifty kings, says the Stranger. The science or art of statesmanship (the Stranger here calls it a *technē* rather than an *epistēmē*), which is exponentially more difficult to master than draughts, will thus be possessed by only one or two or very few men and only they will deserve the name "statesman." In fact, says the Stranger, returning to yet another distinction made near the beginning of the dialogue, namely, that between what something *is* and what it is *called*, only those who possess this *epistēmē* will deserve the name statesman *whether they rule or not*—that is, whether they are in power or not and so whether they are *called* statesman or not, whether they go by the name of statesman in ordinary or everyday discourse or not. Only those with the science of statesmanship really deserve to be called statesmen, whether they are rich or poor, ruling over willing or unwilling subjects, with or without laws, and whether they already go by the name of *statesman* or not (293a).

To illustrate the point, the Stranger returns once again to the analogy between the statesman and the physician—yet another indication that Plato never simply replaces one model by another but is willing to accommodate all kinds of models and images depending on the needs of the dialogue (293b). Whether a physician is following written rules or not, curing us with or without our consent, we call him a physician so long as he is exercising "authority by art or science," that is, so long as he is acting for our benefit, for our health and preservation. This is the only "right definition" (*horon orthon*) of the rule of the physician or of any other ruler, including the statesman, says the Stranger. Similarly, there can be only one right, real form of government (*politeia*), one real constitution or regime, in which "the rulers are found to be truly possessed of science [*alēthōs epistēmonas*]" and do not merely "seem" (*dokountas*) to possess it, and it matters not whether they rule with or without law, over willing or unwilling subjects (293c). Science, *epistēmē*, is the only criterion for determining who is and who is not a true ruler.

Notice, then, that at this point in the dialogue, law, which is often thought to be an essential aspect of statesmanship, has been completely severed from statesmanship, declared completely irrelevant to it. This divide between law and science (the science of statesmanship) will remain intact until the moment, a bit later in the dialogue, when the Stranger redefines the very notion of law by claiming that the true statesman makes *his science his law*—a definition that, as we will see, brings together, through this "metaphorical" understanding of law, two apparently incompatible things (297a). But, well before this moment and this association of science with law, of *technē* with *nomos*, the Stranger here associates the true form of statesmanship as *epistēmē* (or as *technē*) with justice. The Stranger concludes that as long as this statesman is acting with "science and justice" (*epistēmēi kai tōi diakōi*) to preserve and benefit the state, he is acting as the true statesman (293d–e). Justice, then, and not law, is here aligned with the science of statesmanship. The reason for that, we might speculate, because the Stranger does not himself develop or explain this addition, is that justice is something like a law without the law or a law above the law, an unwritten or unspoken law, perhaps, that precedes and trumps all spoken or written laws.

It is at this very moment when science and justice have been brought together that the Stranger draws what seems to be a rather contentious if not positively odious conclusion, so odious, in fact, that it is easy to conclude that Plato is being simply ironic or provocative in order to elicit refutation. He says that as long as this statesman is acting with "science and justice" to benefit the state, it does not matter whether he puts the city on a diet by "killing or banishing some of the citizens . . . by sending out colonies somewhere, as bees swarm from the hive" (293d). The passage might be written off as mere irony or hyperbole, just as the passage in the *Republic* about ridding the state of everyone over the age of ten might be, but there is a very similar passage in the *Laws* that makes this interpretation more difficult to maintain. We looked at this passage earlier in Chapter 3 when we tried to argue, in opposition to Foucault, that the model of the statesman as shepherd never simply disappears and is never simply replaced by the model of the statesman as weaver. In the passage in question, the Athenian argues that the first duty of a good leader, like a good shepherd, is to administer a "purge" (*katharmon*) to separate healthy and well-bred from unhealthy and ill-bred members of the flock, the latter being sent off to other flocks—that is, to other colonies (*Laws* 735b; see also *Republic* 410a). Such a purge or purification is essential, the Athenian explains, because it is a waste of time to expend great efforts on bodies and souls that "nature" (*physis*) and "ill-nurture" (*ponēra trophē*) have come together to ruin and that can in turn ruin a "clean" and "healthy" stock (*Laws* 735b–c). The Athenian admits that this comparison of the city to a herd is just an "illustration" (*paradeigmatos*), but that does not compromise in the least the pertinence of the analogy. In fact, the Athenian says that purging a city is even more important than purging a herd and that the best purge, like all strong "medicine" (*pharmakōn*), is painful and drastic. In this case, it involves exiling or killing those greatest, "incurable" criminals who will harm or pollute the state, something every legislator must do at the beginning or at the founding of a state (*Laws* 735c–736a).[4] Again, it is the model of the statesman as shepherd and not as weaver that provides this important, albeit problematic, trope of purging the state, the flock, of its undesirables, further evidence that none of these models of the statesman (as weaver, as navigator, as physician, or as shepherd) are ever definitively rejected by Plato and all can be redeployed in multiple, strategic ways depending upon the dialogical context.

But now the Stranger adds something else that is troublesome. Rule by science is the only true form of rule, and all other forms, he says, must "be considered not as legitimate or really existent [*ou gnēsias oud' ontōs ousas lekteon*], but as imitating [*memimēmenas*]" this one true form (293e). The better a form of rule imitates this one true form, the better a state is governed, even though, according to the Stranger's rhetoric, *none* of these forms of rule can really be called legitimate or even truly real. Once again, it is difficult not to think of Socrates's critique of writing in the *Phaedrus*. When Socrates contrasts writing with another kind of speech, one that is alive and that has a progenitor, indeed, as we saw in the previous chapter, a father, it is the very same notion of legitimacy that emerges. Socrates speaks of "another kind of speech, or word, which shows itself to be the legitimate brother [*adelphon gnēsion*] of this bastard one, both in the manner of its begetting and in its better and more powerful nature" (276a).[5] Writing is thus opposed in the *Phaedrus* to its legitimate brother, its *gnēsios adelphos*, just as all the non-legitimate forms of rule, forms that are also called *ou gnēsios*, non-legitimate (293e), are opposed to the one true form that all the others must imitate. These bastard or illegitimate forms of rule thus stand in relation to the one true form as writing stands in relation to speech or to what Socrates in the *Phaedrus* calls "writing in the soul." But how are we to understand imitation here? On the one hand, there seem to be better and worse kinds, better and worse imitations of the one true form. On the other hand, all imitations seem to be *mere* imitations, illegitimate and unreal knock-offs of the one true thing—whether that be "writing in the soul" or that genuine form of statesmanship where one or very few men rule by science alone.

The Inferiority—and Necessity—of Written Law

One can imagine what the Young Socrates—or anyone else following the Stranger's argument—must be thinking at this point in the dialogue: Where in the world will they ever find such a statesman to rule with science and without law and how could they, without being true statesmen themselves, ever be sure that such a ruler is not actually a tyrant? They must now confront head-on the question of the propriety and possibility of such a government *without laws* (294a–297a). Though lawmaking seems to belong to the science of kingship, it is better, the Stranger again affirms, "not that the laws be in power, but that the man who is wise [*meta phronēseōs*] and of kingly nature be ruler" (294a). Being ruled by a kingly nature with science, with *epistēmē*, or, here, with *phronēsis*, is far preferable to being ruled by law.[6] The primary reason for this is that law cannot adjust itself to the differences among men and situations. Because nothing in human life is ever at rest, because every individual differs from every other individual and even a single individual differs from himself from one moment to the next, a single law for anything or anyone is always inadequate to the situation. Law is just too simple, it seems, for a complex world in which things are constantly changing. Law is, says the Stranger, like a "stubborn and ignorant man"—a phrase that is again reminiscent of Plato's critique of writing and painting at *Phaedrus* 275d–e. When a genuinely novel situation within the state requires a swift response and radical change, all it can do is repeat what it has already said.

Why make laws, then, if this is the case? Why apply what is so "simple" (*haploun*)—that is, singular, undifferentiated—to "things which are never simple" (294c)? The reason, the Stranger explains, is that the statesman who rules by science cannot sit by each person all his life and he cannot be present to the state as a whole forever. Law is required, quite simply, because the genuine ruler or the originary lawmaker cannot always be present to prescribe what is necessary or appropriate to either the individual or the state at every point in time. It is, in short, the *absence* of the ruler that requires law, the absence of the ruler to each individual, his inability to sit by the side of each citizen, and the eventual withdrawal of the ruler from the entire community—that is, the inevitable death of the ruler or lawmaker. As the Stranger says, returning to the image of the ruler as shepherd, an image that, we recall, Foucault claimed the Stranger simply moves beyond and replaces with the image of the statesman as weaver:

> And so we must believe that the lawmaker who is to watch over [*epistatēsonta*] the herds and maintain justice and the obligation of contracts will never be able, by making laws for all collectively, to provide exactly that which is proper [*to prosēkon*] for each individual. (294e–295a)

As we saw in the previous chapter, only the living speaker (the dialectician, for Plato, the extemporaneous speaker, for Alcidamas) is able to say what is appropriate to the occasion. Writing is unable to take advantage of the situation to produce a desired effect. The Stranger's argument here is very similar with regard to law. The problem with law is that, in its generality, it cannot address the individual, only the majority. Unable to provide "what is proper" for each individual, laws "legislate for the majority [*pollois*] and in a general way only roughly for individuals" (295a). And this is the case whether we are talking about "written laws" or "unwritten traditional customs" (295a). Both legislate only for the majority; both can, at best, only supplement—that is to say, add to and replace—the living word of the originary lawmaker or true statesman who is absent. Both are thus structured by the absence (and, ultimately, the death) of the originary lawmaker, and so both must resort to merely repeating—that is, repeating without variation and without knowledge—past prescriptions, whether written or oral.

The best regime would be one in which laws, whether spoken or, especially, written, are unnecessary and undesirable. It would be a regime in which a genuine ruler or originary lawmaker legislates forever and for every individual.[7] But it was, of course, never possible, not even in the age of the original lawmaker, for the lawmaker or ruler to sit beside each individual, issuing prescriptions throughout these individuals' entire lives.[8] "For how could anyone," asks the Stranger rhetorically, "sit beside each person all his life [*dia biou aei*] and tell him exactly what is proper for him to do? Certainly anyone who really possessed the kingly science, if he were able to do this, would hardly, I imagine, ever put obstacles in his own way by writing what we call laws" (295b). It is clear what is preferable even if it is impossible: rule by a single but omnipresent ruler without law, a ruler or king who would sit by every individual throughout his or her entire life like a personal trainer or shepherd, one's very own life coach, as it were. It is impossible to imagine such a situation. Or, rather, it is possible only to imagine it, in another age, for example, an age in which the God or the Demiurge, acting in concert with other, inferior gods, ruled over all of mankind and all other animals.

Having made a case for the superiority of rule without law, and especially without written law, the Stranger must now give an account of the advantages of written law for all the second best regimes that depend on it. Writing addresses the majority, not the individual, but it at least has the advantage of being able to address citizens through or over time and at a distance. Just as a trainer who is going away for a long time might want to write down his instructions, fearing that they will be forgotten, so the ruler, in order to ensure that there be some rule in his absence and that the ruled might "remember" (*mnēmoneusein*) his instructions, will want to write them down in the form of laws (295c). As in the *Phaedrus*, then, writing or, here, written law functions or at least can function as a reminder, though it must not substitute for living memory itself. It must not, therefore, be valued in its own right but must serve only as a reminder of the living word of the writer himself. The Stranger makes this very clear by imagining a lawmaker or ruler who returns after an absence and wants to revise the laws he left behind or substitute new laws for them to suit the changing circumstances of the state. As the Stranger says, it would be "ridiculous" (*geloion*) if written or unwritten laws about the just and unjust, the honorable and the disgraceful, could not be changed by the scientific lawmaker "for the herds [*agelais*] of men that are tended in their several cities" (295e). Ridiculous it would indeed be if the scientific lawmaker were forced upon his return to adhere to laws he made before his departure that would be not only ineffective but actually detrimental in the current situation. While written laws are thus an aid to human life in times of loss or lack when the originary ruler or lawmaker is absent or dead, they are detrimental to life when the ruler himself is still present or alive.

The Stranger now reminds Young Socrates that people often say—and we are perhaps supposed to recall that Socrates, speaking in the name of the laws, says something similar in the *Crito* (51b–c)—that if one can come up with better laws than the existing ones then one ought either to "persuade" the state to change its laws or else resolve to live in accordance with the old laws. But what, then, about someone who uses not "persuasion but force" (*peithōn . . . biazētai*) to make this change? Returning to his earlier analogy, the Stranger asks about a doctor who, while doing what is best for his patient according to right science, does not persuade his patient to do or undergo something that is contrary to medical prescription or law but actually forces him to do so. Once again, the Stranger suggests, this seems preferable to *not* doing what is best for the patient. So long as the sole criterion for the legitimacy of rule is the science of the ruler, then it does not matter whether the rule is by willing obedience, persuasion, or force. Best of all, of course, as a long passage from *Laws* makes clear, would be to gain the consent of the patient through argument or persuasion (see *Laws* 719e), but force, even without persuasion, seems justified so long as the physician is acting with science. As with the physician, then, no one could criticize the statesman who forces his subjects to do something contrary to written law or oral tradition unless he makes the citizens more unjust and evil rather than better. Again, it would be "ridiculous" for

anyone who was forced to do what was contrary to the law but who nonetheless became better and more just than they were to say that he received
unjust or evil treatment. If health is the *telos* of the physician, and goodness and justice are the end of the statesman, then no one can complain
that he or she has been made healthier or more just through the actions of
the scientific physician or statesman.[9] To return to the terms used earlier,
it does not matter if the physician or the statesman is rich or poor, whether
he is working through persuasion or violence, acting in accordance with law
or not. The "truest criterion of right government, in accordance with which
the wise and good man will govern the affairs of his subjects," is "what is
for the good of the people" (296d–e). For the scientific ruler, the (singular)
end always justifies the (plural) means.

There is thus still a stark contrast between the scientific ruler, present
to his subjects and acting always in their best interest, and law, whether
written or oral. This latter prescribes without being present, without
phronēsis or *epistēmē*, without being able to modify its prescriptions through
an assessment of the situation. All law is, therefore, proto-writing. It shares
all the characteristics of what Socrates condemns in the *Phaedrus*, including, and perhaps especially, its association with distance, with the absence
of the ruler, in a word, with death. For all law, even when simply *spoken*, is
open to iteration, to being repeated in the absence of the speaker—that is,
without his animating and, as we saw Derrida argue, *fatherly* presence. The
relation to *Phaedrus* is clear. If speech is issued by a father who is able to
defend his words, if it is the legitimate offspring of a father who knows the
fertile ground into which his teachings should be planted, if speech is what
gives life and makes fecund, then writing is reduced to mere reminding,
to recollection, where all we can do is "imitate" as best we can that age
when the father/teacher/speaker was present.[10]

Law is here criticized in the *Statesman* in precisely the terms in which
writing is condemned in the *Phaedrus*. Everything that Socrates finds problematic and even dangerous about writing in the *Phaedrus* is echoed in the
Stranger's critique of law in the *Statesman*. When asked a question, the
law—just like writing in the *Phaedrus*—is ignorant and stubbornly repeats
itself, unable to change its response to meet the new situation, cut off from
any living, guiding, or animating presence. We have already seen the close
relationship between the *Statesman* and the *Phaedrus* on the subject of *diairesis*, on the need to divide one's object along its natural joints, on the
necessity of organizing a *logos* like a *zōon*, and, in the previous chapter, on

the question of what kind of discourse is best suited to adapt to changing circumstances in order to be effective. Now we see that the two dialogues are in agreement in their assessment of writing, whether in the form of written law or some other kind of writing: Though written words might look and sound alive, when asked a question "they preserve a solemn silence [*semnōs panu sigai*]"; though they may look "as if they had intelligence [*phronountas*], . . . if you question them, wishing to know about their sayings, they always say only one and the same thing [*hen ti sēmainei monon tauton aei*]" (*Phaedrus* 275d–e). As we saw the Stranger argue earlier, it is discourse, *logos*—discourse in the form of the voice—that is best able to represent its subject, and especially a living being, while writing, like painting, remains inferior, insufficient for the task. These latter are to be used, we recall, only for those who cannot follow a spoken discourse.

Writing is sometimes necessary, perhaps, but it remains inferior to speech or to *logos* at representing a *zōon*. Neither writing nor painting— that is, neither *graphē* nor *zōgraphia*—is appropriate or productive insofar as they remain, precisely, *silent* about the things they depict.[11] Only speech—good speech—is able to answer the questions it is posed and only it is itself organized like a living being—that is, like a *zōon*. Though writing gives the impression of being alive, it cannot come to its own defense, has no power to respond when questioned, and so is left to repeat itself without intelligence—like an ignorant man, indeed like those *zōa* or *thēria* that are characterized as *aloga* (without speech) elsewhere in the dialogues. Hence, speech or, better, "writing in the soul" is better in its origin, which is the soul, and in its nature and capacities, inasmuch as it has the power of the soul, the animating and intelligent breath of the soul, behind it. What is called writing, writing in the restricted sense, is thus opposed to this "writing in the soul," that is, to this "word which is written with intelligence [*met' epistēmēs*] in the mind [*psychēi*] of the learner, which is able [*dunatos*] to defend itself and knows to whom it should speak, and before whom to be silent" (*Phaedrus* 276a). The power or ability that was earlier denied to writing is here granted to "writing in the soul."

We can begin to see the tension—essentially irresolvable—between law (and thus writing) and life. On the one hand, law is always identified with living logos. As Derrida argues in "Plato's Pharmacy," making reference to the way the Laws speak to Socrates in the *Crito*, Plato "always associates speech and law, *logos* and *nomos*, and the law speaks" ("PP" 146). On the other hand, the very essence of law would appear to be its repeatability in the absence of the lawgiver—that is, its *written* character, its relation to the withdrawal, absence, and, ultimately, the death of the writer.

We must return here to the myth of the two ages that we looked at in Chapter 2, for it seems as if the Stranger continues to think according to that schema. The originary lawmaker is present to those over whom he rules just as the speaker is present to the one to whom he speaks or the Demiurge—who is called, as we have seen, a father and a teacher—is present to those over whom he rules during the Age of Kronos. But because the Age of Kronos must one day come to an end, because it has already come to an end, because a speaker cannot always be present to his listener, or a teacher to his disciple, because the lawmaker cannot always be present to all his subjects at the same time and because he must, at some point in time, die, the speaker or lawgiver must consign his speech or teachings to writings, writings that, in the Age of Zeus, can at best only imitate the genuine teachings of the Age of Kronos. It is as if the Age of Kronos corresponds to an age of science or purely spoken law, a mythical age of speech without writing, or, better, following the *Phaedrus*, an age of "writing in the soul," an age in which the father is present and able to adjust and defend his speech, to rule in the present in a direct and immediate way, while the Age of Zeus corresponds to an age of writing or written law, an age in which the father/teacher/lawmaker/demiurge has withdrawn, leaving the universe, precisely, *without* him, so that all mankind can do in his absence is try to recollect as best it can his teachings through the writings he left behind.

The One True Form of Government and Its Six Imitations

It is at this point that the Stranger, again shifting terrain, offers the famous analogy of the state as a ship and the statesman as its captain or its helmsman, an image that should remind us of the myth of the two ages in which Kronos is compared to a helmsman steering the cosmos.[12] The Stranger introduces the analogy by asking this question:

> Just as [*hōsper*] the captain [that is, once again, the *kybernētēs*: see 272e] of a ship keeps watch for what is at any moment for the good of the vessel and the sailors, not by writing rules, but *by making his science his law* [*tēn technēn nomon*], and thus preserves his fellow voyagers, so may not a right government be established in the same way by men who could rule by this principle, making *science* more powerful than the laws? (297a)

To make one's science one's law: this could be, in many ways, the watchword of the entire dialogue. It at once expresses the goal or *telos* of science (either *epistēmē* or *technē*) and exposes the inherent contradiction or conflict between science and law. While such a science, as we have seen, would be perfectly adaptable, infinitely flexible, able to adapt to every changing situation, able to take advantage of the opportune moment, the *kairos*, law would always remain to some extent inflexible—that is, structurally and not simply by accident rigid and difficult to change, related always to the loss or absence of the ruler, the loss of a father who can defend or change it. If the essence of all law, even spoken law, is defined by its "written" character—that is, by its relation to the absence or death of the ruler, lawgiver, or writer, to memory as a mere reminder—then *epistēmē* or *technē*, and so statesmanship properly speaking, is defined by its relation to speech or to "writing in the soul," to an even more immediate form of communication, to presence and to life. In the end, *epistēmē* and written law are about as different as the Ages of Kronos and Zeus. In order for the law to appear at all, it must be repeatable or iterable at another time, repeatable in a context for which it was not destined, iterable in the absence of any speaker. If the very structure of law (whether written or oral) is "written" in this way—that is, able to be repeated in other contexts and in the absence of the father—then to make one's science one's law is to dream of a law that is infinitely adaptable and is never simply repeated, a sort of *natural law* or *living script*, an impossible law or a law before the law.[13]

The true statesman is the one who makes his science his law, who rules without written law and so rules with wisdom or knowledge and justice. He rules the state like a captain or a helmsman rules a ship, or like reason rules the body or the body politic. In the *Laws*, it is *nous*, in fact, that is itself called a pilot or a helmsman who, "when combined with senses," that is, when combined with the sailors, "will accord salvation to ships in stormy weather and calm" (*Laws* 961e). Appealing in the *Statesman*, as in the *Laws*, to a soteriological discourse on the state's "salvation," the Stranger says that so long as the rulers preserve (*sōzein*) the citizens by dispensing absolute wisdom and justice in order to make them better, then they will be committing no error. As the Stranger has said, very few men could ever possess this science, this "political science," so as to administer a state with wisdom (297c). There is thus only *one* right form of government and it is to be found "in some small number of person or one person"; all other forms of government are better or worse "imitations" (*mimoumenas*) of this one true form.[14] Once again, unsurprisingly, speech, wisdom, and science, *epistēmē*, are all on the side of the one—and on the side of life—while writing, law, and imitation are on the side of the many—and death. The imitating forms are thus "not perfectly right" (*ouk orthotaton on*) (297d), for they all contain an "error" (*hamartēma*) or a missing of the mark with regard to that one true form.[15] While only a God or a god-like man could *really* save or preserve the state, those who live in the Age of Zeus where there is no such God and where god-like men are extremely rare and difficult to recognize must rely on the supplement of written law. They must, therefore, as a second choice and in order for their states to be "preserved" (*sōzesthai*) so far as possible, employ written law and prohibit any transgressions of that law. While law, written law, is always a distant second best, when no true statesman can be found this second-best law must be not only allowed but strictly obeyed. The next best state after the scientific one where the question of law, whether written or unwritten, is irrelevant, is thus, the Stranger argues, a state in which strict obedience to the law is required.[16]

The Stranger describes how these second choices come about by returning to the previous "images" (*eikonas*)—which, I note again, Plato never abandons or rejects—of the physician and the captain of the ship (297e).[17] Imagine, says the Stranger, patients of a doctor or passengers of a ship complaining about the abuses, about the unequal, arbitrary, and unjust treatment, they have received at the hands of the physician or pilot and so deciding to overthrow their rule (298a). Imagine, he says, that they then call an "assembly" and allow everyone, the "multitude" without skill or art or any special qualification, amateurs (*idiōtai*) all, to offer their opinions about disease and navigation. Imagine that they then adopt written laws for healing or navigating based on these opinions of the majority. Young Socrates sees that it would clearly be "absurd" for the unskilled or lay multitude to decide on medical matters or navigational ones in this way, writing laws on tablets or slabs or adopting unwritten ancestral customs so that "navigation and the care of the sick should [forever] be conducted in accordance with these provisions" (298e). But imagine further, the Stranger says, that the rulers are elected annually by lot (a practice common in Athens)—either from the rich or among the multitude—and that these rulers must agree to rule in accordance with previously determined laws and customs, with those who disobey being punished accordingly (298e). Moreover, the conceit goes on, if anyone is found to be investigating the arts of medicine or navigation they will be called not a physician or ship captain but a "stargazer" (*meteōrologon*) and a kind of "loquacious sophist" (299b–c)—terms that recall, of course, the two *informal* charges against Socrates, namely, that he was a nature philosopher and a sophist. Such a person may thus be haled into court for—and now there is an echo of one of the *formal* charges against Socrates—"corrupting the young" and "persuading" them to question the laws of medicine and navigation "according to their own will." If found guilty of persuading against the laws, they will then be sentenced to death, for they will say that "nothing ought to be wiser than the laws," laws that are determined, recall, by the artless majority but that are public and available for all to learn and obey (299c).

The Stranger clearly sees the dangerous consequences of this argument. If such written rules were applied to all the arts—to hunting, painting, all kinds of imitation, carpentry, husbandry, herdsmanship, prophecy, draught-playing, arithmetic, geometry, and so on—that is, if these arts were practiced "by written rules and not by knowledge [*kata technēn*]," the arts would be utterly ruined (299d). With all investigation prohibited, the work of the arts would be destroyed and life itself, as Young Socrates too sees, would become "unendurable" (*abiōtos*) that is, untenable, unsustainable, unlivable.

And yet, as unlivable as such a life may be, there is, says the Stranger, an even worse possibility: If someone elected to watch over these arts in accordance with written rules were to act out of desire for some personal gain against the laws and without any knowledge, this would result in an even worse evil than simply following the written law (300a). For at least the laws will have been made only after "long experience [*peiras*]" and consideration and only after persuading the people.[18] To act contrary to these laws would thus be a greater error than adhering strictly to them. For these laws, written by men "who know in so far as knowledge is possible, are imitations [*mimēmata*] in each instance of some part of truth" (300c). While imitation is still imitation and not the truth, it is, here, at least still an imitation *of the truth* and not, as we saw early, a *mere* imitation with no truth in it. That is how the Stranger can maintain at this point in the dialogue two distinct but not contradictory claims: First, there is a single science of statecraft that has no need of law and, second, one should follow law in the absence of this science and prescribe penalties for innovation. In other words, it is best to live in a regime without law but, in the absence of such a regime, laws must be made and strictly obeyed. Following written laws without knowledge is worse than following an unwritten art or science, worse than the rule of someone whose science is his law. But going against written laws without knowledge and without the experience and consideration of those who made the laws is worse than simply following those laws. While the true statesman rules by his science or his art, something of which the multitude is incapable, the multitude is at least able to follow established law and it is able, in some limited way, to help reform law on the basis of experience.

The Stranger recalls how the "real statesman" would be able by his art to change laws when he thought it best by sending new laws to his absent subjects. But when others try to change the laws—in imitation of the true statesman who changes his own written laws when he sees that conditions have changed—they would be trying to "imitate reality [*mimeisthai . . . alēthes*]" but "they would be imitating badly" (300d). In absence of the best regime (the one without written law and in accordance with science and knowledge), one must accept what is second best (rule by a majority in accordance with laws that have been drafted after experience and consideration), so as to avoid the worst (the rule of those who go against the law without science or knowledge). Writing in the Age of Zeus is thus necessary, but it is always only second best, a supplement that tries to imitate what is truly best, the one true art of statesmanship that no actual statesman is perhaps ever able to exercise or embody. When the best regime and the best form of life is impossible to achieve, one must accept living with the laws and the lesser form of life that comes with that.

The Stranger's concludes: "Such states . . . if they are to imitate well [*mimēsesthai*], so far as possible, that true form of government [*politeian*]—by a single ruler who rules with science [*meta technēs*]—must never do anything in contravention of their existing written laws and ancestral customs" (300e–301a). The best form of rule in the Age of Zeus is thus one in which a single ruler "rules according to laws and imitates the scientific ruler," a ruler who will therefore be *called* a king (*basileus*), "making no distinction in name," says the Stranger, "between the single ruler who rules by science and him who rules by right opinion if they both rule in accordance with laws" (301a–b). Though this latter rules only by "opinion" (*doxēs*) and not knowledge, he will have the same name, *king*, as he who rules according to science. It is as if names too could be imitations of their true form, traces, as it were, of themselves (301a). The name *king* is appropriate to both because both, according to the Stranger, rule "in accordance with laws," a formulation that nicely elides the difference between

these two kinds of "law," the *one* law that coincides with the science of the
true statesman and the *multiple* laws of the one who, in some still unexplained way, imitates him.

As for the other forms, when the rich imitate the true form of government by following written laws (but without knowledge), then this is called
an aristocracy, and when they rule without regard for written rules it is
called oligarchy. The Stranger can thus say that the "five names of what
are now called the forms of government have become one" (301b).[19] That
is because Tyranny, Oligarchy, Democracy (as rule by the majority or rule
by the mob), Aristocracy, and Kingship (without knowledge and following law) are all *imitations* (from worst to best) of the one true form of government, imitations of the one true Kingship. But this means that even
mob democracies and oligarchies, even tyrannies, are in some way imitations of the one true form of statesmanship. Even the tyrant resembles
the true statesman or king insofar as he is a single ruler who does not follow the law; he differs from him, of course, insofar as his "imitation is inspired by desire and ignorance [*epithumia kai agnoia*]," that is, by personal
gain rather than the well-being of the state (301c). We are back to the two
versions of the statesman as shepherd in Book 1 of the *Republic*, the one
who cares for the herd for his own benefit, as Thrasymachus maintained,
and the one who, as Socrates counters, cares for the herd for the sake of
the herd.

Because the people seem to believe that there is no single person who
could rule by the perfect form of government, they must instead come
together to make written laws. Because they do not believe that there is
someone "worthy of such power or willing and able by ruling with virtue
and knowledge to dispense justice and equity to all," someone who would
be, "like the ruler of bees in their hives, by birth [*emphuetai*] pre-eminently
fitted from the beginning in body and mind"—language that is echoed in
the *Politics* (see, for example, Book 7, 1325b)—they must instead "follow in
the track [*ichnē*] of the perfect and true form of government by coming together and making written laws" (301d–e). It is unclear whether the emphasis is being placed here on the extreme rarity of a true statesman or on the
inability of the multitude to recognize him, but what is clear is the value
the Stranger puts on such a ruler. If there were such a person to direct
the people "to their weal [*eudaimonōs*]" and the people were able to recognize
him, then they would undoubtedly follow him. In his absence, however,
they must follow his "track," in his trace, and imitate this one true form,

as if, following a certain Plotinus, form were always the trace (*ichnos*) of the formless, as if the forms of government in the Age of Zeus were traces of that formless form of rule that typifies the Age of Kronos.[20]

It is no wonder, then, says the Stranger, that states have such problems when they are built on such a foundation—that is, without knowledge and with written laws. As they earlier agreed, such a reliance on law would have ruined any other art; the real wonder, then, is that there is any "stability" (*ischuron*) at all in states. While many states have, of course, been shipwrecked and destroyed by men who *thought* they knew the art of statesmanship but were, in fact, wholly ignorant of it, some states have survived and have not been overthrown (301e–302b). Though the Stranger does not emphasize the point, knowing one does not know—that is, in this case, knowing that one lacks the true science of statesmanship—is far better than thinking one knows it when one does not.

It is their duty, then, says the Stranger, to determine which of these "not right" forms of government are easiest to live with and which most difficult (302b). The Stranger proceeds to divide the three forms (rule by one, few, or many) into six by distinguishing a form that follows and applies the law from one that does not (302d). They had begun this earlier by distinguishing Royalty from Tyranny and Aristocracy from Oligarchy. Now they need to divide the name Democracy, which is "already twofold in meaning," into legitimate democracy or rule by the multitude and mob rule, so as to get six forms, all of which are to be distinguished from the one true form.

While the distinction between rule by law and rule without law was not useful earlier (the essential distinction was between ruling *with* science and ruling *without*), the distinction now neatly bisects the six forms—which are all to be distinguished from the one true, scientific form, the seventh, as it were, which is beyond all established and stable law and yet remains a model for that law. When monarchy is "bound by good written rules, which we call laws, it is the best of the six; but without law it is hard and most oppressive to live with" (302e). Rule by one is thus both the easiest and the hardest to live with, for this can be either Kingship or Tyranny depending upon whether the rule is with or without law. Government of the few is intermediate in both good and evil, just as few is intermediate between one and many (303a). As for rule by the multitude, the fact that power is shared out among so many means that it is able to do little good or evil. As a result, in an orderly, lawful government Democracy is the worst regime to live under and, in a lawless regime, one "without restraint [*akolastōn*]," it is the best (303a–b).

The Stranger concludes that life in a kingship or royalty is best of all— with the exception, of course, of the seventh, the one true form, "for that must be set apart from all the others, as God is set apart [*ekkriteon*] from men" (303b). Such a life is set apart but also, it seems, selected, elected, and elevated, for that is the life whose tracks or traces we must follow. The re-lation between the best form of government and life and the others seems to parallel the relation between God and men, and, perhaps, the Age of Kro-nos and the Age of Zeus. It is a form of rule and life that is set apart and yet must be imitated through what I called earlier a sort of *political anamnesis*.

Though this "single ruler who rules with science" cannot be said to be the *same* as the Demiurge or the divine ruler in the Age of Kronos, he certainly seems to be positioned in the same place—that is, as a model or ideal to be imitated or recollected. The six forms of government—that is, in order from best to worst, kingship, aristocracy, law-abiding democracy, mob democracy, oligarchy, and tyranny—are all *imitations* of the one true form of government. But now there comes a claim about the nature of imi-tation that recalls the Stranger's earlier claim that these imitative forms should be "considered not as legitimate or really existent, but as imitating" this one true form (293e). While the Stranger seems to have been arguing that the six imitative regimes are better or worse to the extent that they imitate the best state or follow its tracks, he now says that all non-scientific rulers are not statesmen at all but "partisans" (*stasiastikous*) who "preside over the greatest counterfeits [*eidōlōn*]"; as such, "they are themselves counterfeits, and since they are the greatest of imitators [*mimētas*] and cheats [*goētas*], they are the greatest of all sophists" (303c). This would seem to suggest that all these imitations of the truth are, once again, *mere* imitations—that is, simulations or counterfeits—and that there are, as a result, *no* statesmen among men, not even bad ones, only sophists.

As the Young Socrates notices, the object of their previous search, the sophist, has cropped back up in their present search. It looks as if all states-men in the Age of Zeus are no better than sophists, while the only true statesman is the impossible one, the one who makes his science his law, like God or a god in the Age of Kronos. Perhaps, the Stranger suggests, if they are able to distinguish more clearly the statesman's art they will be able to save *some* of these imitators from the category of cheats and charlatans.

The Statesman as Weaver

But now another class appears in close proximity to the statesman, one even more difficult to separate off and differentiate because more "akin" (*syngenes*) (303c–d). Like refiners of gold, says the Stranger, they must, having separated off the earth and stones, use fire to remove those things that are more "akin" to gold, that is, the copper, silver, and adamant, until only unalloyed gold remains. They have already separated off everything that is "different and alien and incompatible" with the science of statesmanship—that is, all those who are *called* statesman in most states, all those who *seem* akin to the statesman, namely, kings, aristocrats, democrats, oligarchs, and tyrants, all those who are not really statesmen at all but sophists. Now they must consider only what is "precious [*ta timia*] and akin to [*xyngenē*] statesmanship" (303e).

What remains, then, are the arts of the general, the judge, and the orator or rhetorician. To eliminate these things so "akin" to the statesman and so show him "alone and by himself and undisguised" (304a), the Stranger distinguishes sciences such as "music" and the "sciences of handicraft [*cheirotechnias*]," that is, sciences that are learned, from the "science which decides whether we ought to learn [these sciences] or not" (304c). There are, then, on the one hand, the *many* different sciences to be learned (including the sciences of the hand) and the *one* science that decides what sciences are to be learned and when (a science of the head, perhaps). The difference here is between knowing something and knowing whether that something is to be learned, between knowing *how* to do something and knowing *whether* or *when* to do it.

It is this distinction that allows the Stranger to distinguish oratory from statesmanship. The Stranger can at once admit that the orator "partakes of the kingly art because it persuades [*peithousa*] men to justice and helps to steer the ship of state" (304a) and yet still distinguish that art from statesmanship, "the science which decides whether to persuade or not should control that which can persuade" (304c). It is in this way that oratory can be integrated into the state and yet put in its place, rehabilitated or reformed, just as it is in *Phaedrus*. While the orator or the rhetorician is thus not the statesman and rhetoric is not statesmanship or lawmaking, we can imagine the art of oratory lending a hand to statesmanship by helping to craft, as the Athenian suggests in the laws, persuasive preludes to the law. If the orator is able to take advantage of the opportune moment to persuade an audience, the statesman decides whether it is the opportune moment to take advantage of that opportune moment.

Given, as we saw in Chapter 2, Plato's emphasis on teaching in the myth of the two ages and in his criticism of Heraclitus and the Heracliteans, it is not insignificant that rhetoric is here defined as a science that has the "power of persuading a multitude or mob by telling edifying stories [*mythologias*], not by teaching [*didachēs*]" (304c–d). True statesmanship, as opposed to rhetoric, is able to teach and not just persuade; as such, it is the science that "controls [or rules over] the sciences of persuasion [*peistikēs*] and speech [*lektikēs*]." It is statesmanship, therefore, that decides what is and is not to be done, whether by "persuasion [*peithous*] or"—for this possibility has not been definitively excluded—"some exercise of force [*bias*]" (304d). This is precisely what we saw earlier in the example of the physician who is acting as a physician only insofar as his actions aim at health, regardless of whether he uses persuasion or force. It is the statesman who knows the right moment, the *kairos*, for using speech, while the orator knows the right moment, the *kairos*, for persuading the multitude in that speech. Rhetoric is thus separated off from statesmanship as a "species" (*eidos*) subservient to it (304d–e)—yet another similarity between *Statesman* and *Phaedrus*, where dialectics is defined as the true art of rhetoric and anything beyond dialectics is considered to be a mere supplement to that one true art (see *Phaedrus* 265d–e). Once again, we see Plato usurping the privilege of the orator, here by the statesman, in the *Phaedrus* by the dialectician—a parallel that should cause us to wonder whether there really could have been a third dialogue called the *Philosopher* insofar as the true sophist and/ or true statesman is perhaps also a philosopher.

Having separated off the orator from the statesman, they must now separate off the general. The "function or power [*dunameōs*]" of determining *how* war is to be waged is here unbegrudgingly attributed to the art of generalship. Statesmanship, for its part, is the science of deliberating and determining *whether* to go to war, *whether* to make war or peace. Hence the art of generalship is also "subservient" (*hypēretikēn*) to the art of statesmanship. Only the kingly art is bold enough, says the Stranger, to declare itself "mistress" of the art of war (305a), and the same is true for the statesman when faced with the "power or function" of judges. While judges must judge "contracts" in as impartial a way as possible, unmoved by bribes, fear or pity, enmity or friendship, they are simply guardians of the law made by the statesman and so are servants of the kingly art. The statesman is thus not an orator, a general, or a judge, but the one who must rule over or control these three. As the Stranger says, the truly kingly art "ought not to act itself [*ouk autēn dei prattein*]" but should "rule over the arts that have the power of action [*archein tōn dunamenōn prattein*]; it should decide upon the right and wrong time [*enkairias . . . akairias*] for the initiation of the most important measures in the state, and the other arts should perform its behests" (305d).

The statesman is a master of the art of measurement in accordance with the mean, the one who determines not how to do something but whether it is proper, opportune, or fitting to do it. All the other arts have a "peculiar sphere of action" and thus have special names corresponding to those actions, while statesmanship is the art that determines when and for what purpose these arts should be deployed. The one art that oversees all the others, that rules over the others, that watches over the laws and all the other things in the state and "weaves them all most perfectly together," is thus the one art that deserves the name "statecraft" (*politikēn*) (305e).

Using the paradigm or "model" provided earlier by weaving, they can now describe the "kingly process of weaving [*basilikēn symplokēn*]," "its nature, the manner in which it combines the threads, and the kind of web it produces" (306a). In short, the Stranger says, the art of the statesman consists in blending or weaving together different kinds or parts of virtue in the state, especially when "one part of virtue is in a way at variance with another sort of virtue" (306a).[21] For, contrary to popular opinion, certain parts of virtue—while remaining virtuous—are often not friendly to one another but hostile and opposed (306b). The Stranger gives the example of opposite qualities like quickness and slowness: *both* are often praised in particular cases, the first as energetic or courageous, the second as gentle or restrained, even when the men who possess these different qualities are at odds with one another. It is only when things are *too* quick or sharp for the occasion, for the *kairos*, when they "seem to us out of place [*akaira*]," that they are called "violent or mad"; only when they are *too* slow or heavy that they are called "cowardly and sluggish" (307b–c).[22] Hence blame is attached to an act not because it is fast or slow but because it is too fast or slow *for the occasion*. Once again, the art of statesmanship is the art of the mean, an art that aims at what is appropriate, proper, fit or opportune, an art that is concerned not with relative measure but with praiseworthy or blameworthy action in accordance with the mean. It is an art of the mean that oversees all the other arts, which are also, of course, as arts, arts of the mean.

While the courageous and the self-restrained can thus both be praised, they are often, says the Stranger, "like two parties arrayed in hostility to each other," unable or unwilling to mix with one another (307c). Moreover, because such men tend to praise the qualities akin to their own and blame the ones that are foreign and opposite, like will invariably gravitate to like and even greater enmity will arise between the two sides (307d). Without the intervention of the statesman, this enmity can thus become a terrible "disease" (*vosos*) that will end up affecting "the whole course of life" (*tēn tou zēn paraskeuēn*) of the state (307e).[23] For the self-restrained or moderate will desire to live a quiet, peaceful life, minding their own business, both in private affairs and with regard to other states, and, though virtuous in its own way, this desire will often prove to be "inopportune" (*akairoteron*) and excessive and will eventually make those who have this virtue unwarlike, themselves as well as their children, causing them to fall prey to their enemies and so "pass by imperceptible degrees from freedom to slavery" (308a). The courageous, for their part, in their "excessive desire for a warlike life" (308a), will often get their states involved in wars that will destroy the state and turn them, just as those of the opposite virtue did, into slaves. Hence there is a constant enmity between these two classes of virtuous men—between *doves* and *hawks*, as it were. While both are important parts of virtue, while both are virtues, they are by nature at variance with one another and, left to themselves, each will destroy the state. The best state, and the best life to which it gives rise, will have to seek the mean between these extremes.

The statesman, like the practitioner of every "constructive science," will thus bring together the best elements at his disposal for the good of his art; he will bring together good men, albeit of different virtues, and, after testing them in play, turn them over to those able to teach and supervise them. Just as the art of weaving, as we saw, supervises and commands the carders and others who prepare the materials for the web (*symplokē*), so the statesman—by means of the kingly art (*basilikē*)—exercises the function of supervising the other arts, admitting only those able to produce "characters suitable [*prepon*] to the constitution" (308e). Once these characters of the courageous and self-restrained have been selected and prepared, they are woven together, says the Stranger in accordance with the "simile" (*eikona*) of weaving, like warp and woof, with the coarser and stronger material of the warp (the courageous character) being woven together with the softer and more flexible material of the woof (the self-restrained character) (309b).[24]

It thus looks like the statesman is indeed much more a weaver than a shepherd. And yet, the Stranger now insists, those who have no capacity at all for courage or self-restraint—the extremes of virtue that statesman-ship tries to promote and combine—those who, "by the force of an evil nature, are carried away into godlessness, violence and injustice [*eis atheotēta kai hybrin kai adikian*]," should be removed, either by death or exile or by having their rights in the city revoked. This looks very much like that "purge" brought about by the shepherd in the *Laws*, which suggests yet again that the model of the statesman as shepherd has not been completely replaced by that of the weaver.[25] The statesman is a weaver/shepherd/nav-igator/physician: There is no need to choose insofar as the statesman ex-ceeds each of these designations.

The statesman's art weaves together different men of different virtues. But it also weaves or brings together, within each individual soul, two opposed and seemingly incompatible elements, the divine and the human, since the soul is itself a combination of these two elements: "first it binds the eternal part of their souls with a divine bond, to which that part is akin [*prōton men kata to xyngenes to aeigenes on tēs psychēs autōn meros theiōi xynarmosamenē desmōi*], and after the divine it binds the animal part of them with human bonds [*meta de to theion to zōogenes autōn authis anthrōpinois*]" (309c). The divine is that "really true and assured [*bebaiōseōs*] opinion about honor, justice, goodness [*tēn tōn kalōn kai dikaiōn peri kai agathōn*] and their opposites," for when this opinion emerges in men's souls, says the Stranger, it does so in a "godlike race [*daimoniōi . . . genei*]" (309c). Though the princi-ple of "like to like" remains intact (the divine bond to the divine part of the soul, human bonds to the human part), the human soul is itself a com-bination, a weaving together, of things that are opposed, the divine—which is eternal [*to aeigenes*]—and the animal [*to zōogenes*], the living animal, which is always, as we saw, mortal, *thnētos*. That which is everlasting, *aei*, is thus combined with that which is mortal or creaturely. The (single) divine bond is woven together with (multiple) human bonds. It is as if the Age of Kro-nos, and the single form of rulership to which it corresponds, were being thought together with—woven together with—the Age of Zeus and its multiple forms of rulership.

This power to combine not just opposite virtues but the human and the divine belongs only to the statesman (*politikos*) and good law-giver (*agathos nomothetēs*). Only he, guided by the kingly art or "by the inspiration of the kingly art [*basilikēs mousēi*]," can implant this "true opinion" in a soul that has been rightly educated. It is this opinion, this true opinion, that makes the courageous soul "gentle" and "ready to partake of justice," instead of "inclin[ing] more towards brutality [*pros thēriōdē tina physin*]," that is, toward the animal (309e). As for the "decorous nature," it becomes "truly self-restrained and wise with this opinion," rather than a mere "simpleton" (*eutheias*) (309e). True opinion—which partakes of the divine—is that which tames the brute, the *thērion*, in man and makes the self-restrained more wise.

It is right at this point, very close to the end of the dialogue, that the Stranger uses a word to describe law that Socrates in the *Phaedrus* uses to describe writing, that famous word *pharmakon* around which, as we saw, Derrida weaves the quasi-totality of "Plato's Pharmacy." It is, says the Stranger, only in those "who were of noble nature from their birth and have been nurtured as befits such natures" that this opinion "is implanted by the laws," as "the medicine [*pharmakon*] prescribed by science [*technēi*]" to produce a bond that is even more divine, a bond "which unites unlike and divergent parts of virtue" (310a). Right opinion is thus the ultimate *pharmakon*, the greatest remedy, administered by *technē* to bind together opposite virtues. It is a kind of *supplement* that improves, as it were, on human nature itself or, better, that brings out the divine part of human nature. It is a kind of supplement within the Age of Zeus that brings us back—or brings us as close as possible—to that mythical Age of Kronos. Whereas human nature divides up virtues into different natural types, right opinion is able to bring them (back) together. But right opinion is also, like a piece of writing, what must be retained and respected in the *absence* of genuine knowledge or the one true statesman. Hence, the statesman weaves together the human and the divine as if he were weaving traces of the Age of Kronos into the Age of Zeus. And he does this, notice, as in the *Laws*, within the human soul, because there is within the human soul a part that is akin to the divine as well as one that is akin to the animal.

Once the divine bond, this singular bond of "true opinion," has been established, it is easy, says the Stranger, to devise the multiple human bonds to follow—that is, laws and customs regulating marriages and adoptions between states and marriages and the procreation of children within the state (310a). Rather than considering such things as money and power when it comes to marriage; rather than having the family as one's "chief care" (*epimeleian*) by seeking spouses and wanting to have children that are most like oneself; rather than seeking "natures of their own kind," both "classes" (*ta genē*)—the same word, notice, that was used at 271d to describe the various animal groupings in the Age of Kronos—must be encouraged to seek mates of an opposite character for themselves and their children (310b–c).[26] For "with no admixture of a self-restrained nature," courage, which begins strong and flourishing, ends up in "madness" (*maniais*), while the modest or self-restrained soul, without being blended with courage, ends up sluggish and crippled (310d–e).[27]

These "human bonds" can easily be created, says the Stranger, as long as both classes have "one and the same opinion about the honorable and the good." That is, so long as both classes have the same "divine bond," the same right opinion, about justice and honor, both will agree to accept these laws of marriage and generation. Hence the art of the statesman consists in weaving together the self-restrained and the courageous characters "by common beliefs and honors and dishonors and opinions and interchanges of pledges," thereby producing a smooth and well-woven fabric and "entrusting to them in common for ever [*aei*] the offices of the state" (310e–311a). "For ever," notice, *aei*, just as dialectic in the *Phaedrus* is said to be capable of yielding seeds in other minds and of "continuing the process for ever [*aei athanaton*]," seeds that "make their possessor happy, to the furthest possible limit of human happiness" (*Phaedrus* 277a).

All this will be done, the Stranger now argues in conclusion, by choosing a single leader or a supervisor with both qualities, and by choosing boards or councils with men of both classes, officials who are self-restrained (careful, just, and conservative) and courageous (bold and capable of action) (311a). Both qualities must be present for a state to be "entirely prosperous in public and private matters" (311b). This weaving of characters and binding together of the human and the divine is the ultimate "end," the *telos*, of the statesman's activity (311b).[28]

Having thus "clothed" the state and its inhabitants—both slaves and freemen—in this fabric, the statesman then rules and "watches [*epistatēi*] over them" (311c), says the Stranger, using the same verb (*epistateō*) used earlier to describe how the gods watched over man in the Age of Kronos (see 271e, 294e). The statesman's art is indeed comparable to the art of the weaver, but it is also an art that tries to imitate the one true form of statesmanship that has been associated with the Age of Kronos, the one true form where one's science is one's law, the one true form where the ruler is just as much a shepherd as a weaver. Because the statesman in the Age of Zeus is not the Demiurge or the God of the Age of Kronos, because law always entails the absence and, ultimately, the death of the lawmaker, the statesman must rely upon the supplement of law in the form of right opinion, the divine bond, and the many human bonds to order and regulate the state. The statesman or the one with the kingly art thus clothes man with the garments that are most proper to him, the laws and customs and opinions that weave the individual into the fabric of the state. And he does this in imitation of and in order to come as close as possible to this state of nature that would have been the Age of Kronos. The statesman thus weaves the *supplement* of clothes, as in the myth told in the *Protagoras*, so that man can begin to resemble that state of affairs in which he had no clothes but nonetheless lived peacefully with himself and other animals.

The *telos* or end of statesmanship is thus a weaving together of opposite characters in the state and a bringing together of human and divine for the sake of the common good or, better, for the sake of *life*. Just lines before the final exchange of the dialogue in which the Young Socrates thanks the Stranger for "a most complete and admirable treatment of the king [*basilikon*] and statesman [*politikon*]" (311c), the Stranger recalls, one last time, the theme of *life* that we have been tracking throughout this work. The *telos* or goal of the "kingly science [*hē basilikē technē*]," he says, is to bring these various characters together "by friendship and community of sentiment into a common life" (*hopotan homonoiai kai philiai koinon xynagagousa autōn ton bion*)—that is, a common *bios* or way of life (311b–c). Weaving together various elements of the state into a single fabric is all done in the name or with the goal of a common life. In the end, life, human life, is the proper object or end of the statesman. While the theme has been, as it were, woven throughout the entire dialogue up to this point, it here appears, right at the end, in its most explicit form and under its own name, or at least one of its names—that is, under the name of *bios*.

CHAPTER 7

Plato and the Invention of Life Itself

Well, but on your own view, life is strange [*deinos ho bios*]. For I tell
you I should not wonder if Euripides' words were true, when he says:
"Who knoweth if to live is to be dead [*to zēn . . . katthanein*] / And to
be dead, to live [*to katthanein de zēn*]?"

—SOCRATES TO CALLICLES (*Gorgias* 492e)

The entire range of the inflections of the verb "sein" is determined
by three different stems. The first two stems to be named are
Indo-European and also occur in the Greek and Latin words for
"being." The oldest, the actual radical word is *es*, Sanskrit *asus*, life,
the living, that which from out of itself stands and which moves and
rests in itself: the self-standing [*Eigenständig*].

—MARTIN HEIDEGGER[1]

The *Statesman* ends by defining statesmanship as the "kingly science"
(*hē basilikē technē*) whose goal, or *telos*, is to weave together various characters "by friendship and community of sentiment into a common life," a
common *bios* (311b–c). But what exactly is life, life as either *bios* or *zōē*, in
Plato? This is the question we have been circling around throughout
this work and that we must now take on more directly. Though it will be
impossible to do justice here to this enormously difficult question, we
will at least begin to see a difference develop between *zōē* and *bios* that is
never explicitly argued for but whose profile can be seen haunting Plato's dialogues, a difference that, at least initially, looks very much like the
difference between human, political life and what has come to be known
as bare life but that is quickly transformed into the difference between
what is called life and real life or *life itself*. Let me begin to show, then, how
the question of life itself, the question of what makes life life and what
makes what we *call* life a lesser version of life itself, begins to creep into

the Platonic corpus, making it possible for someone like Plotinus to speak some six centuries later not simply of beings that are alive but of "real being" that "must necessarily be in life, and in perfect life."[2] As I will argue, it is this development and transformation that is most significant in Plato and most original to him, a transformation that will be taken up in various ways in Western Philosophy from Plotinus to Hegel and beyond.

To begin to show how a thinking of life as life itself paves the way for someone like Plotinus to speak of a true, perfect, or real life that leaves *traces* of itself in everything that is *called* life, we need to distinguish between two terms for life, *bios* and *zōē*, and their verbal analogues in the dialogues. The most striking and commented upon contemporary text on this subject is, of course, Giorgio Agamben's *Homo Sacer*, which claims on its opening page:

> The Greeks had no single term to express what we mean by the word "life." They used two terms that, although traceable to a common etymological root, are semantically and morphologically distinct: *zōē*, which expressed the simple fact of living common to all living beings (animals, men, or gods), and *bios*, which indicated the form or way of living proper to an individual or a group. When Plato mentions three kinds of life in the *Philebus*, and when Aristotle distinguishes the contemplative life of the philosopher (*bios theōrētikos*) from the life of pleasure (*bios apolaustikos*) and the political life (*bios politikos*) in the *Nicomachean Ethics*, neither philosopher would ever have used the term *zōē* (which in Greek, significantly enough, lacks a plural). This follows from the simple fact that what was at issue for both thinkers was not at all simple natural life but rather a qualified life, a particular way of life.[3]

While this distinction is not as clear as Agamben presents it and, in Plato at least, there are many places where the two words are used more or less interchangeably, this distinction *can* be found in Plato to some extent and, as we will see, it makes possible not just a concept of mere or bare life in relation to political life but, more importantly, the concept of *real life* or *life itself.* In order to demonstrate this, however, we need to look at how Plato uses these different words from very early dialogues on and how, especially in later dialogues, a notion of *zōē* as real life or life itself begins to become detached from *bios* or anything that might be *called* life.

Bios *and Human Life*

Let me begin, then, with *bios*, a word that is much more common in the dialogues than *zōē*. *Bios* names, first of all in Plato, a way of life or a kind of life, a life that can be better or worse depending on how it is lived. Perhaps the best-known use of the word is to be found in the *Apology*, where Socrates famously says that "the unexamined life is not worth living" (*ho . . . anexetastos bios ou biōtos anthrōpōi*)—that is, that the unexamined *bios* is not *biōtos* (*Apology* 38a).[4] *Bios* means here, clearly, a way of life, a certain kind or quality of life, a life that can be compared to other lives, the unexamined life as opposed to the examined life, or, elsewhere, the life of pleasure as opposed to the life of the mind. Agamben is right to recall the *Philebus* in this regard insofar as nineteen of the eighty-five uses of the term *bios* in the Platonic corpus are found in this dialogue, with the *Republic* coming in a very close second with eighteen. The *Phibelus* is in essence an inquiry into the relative merits of the life, the *bios*, of pleasure, the life or *bios* of the mind, and the life that combines the two. The comparison is between three different *bioi*, then, three different lives or ways of living for men, or indeed for other *zōa*, other living beings, because Socrates says at one point in the dialogue that a life or a *bios* of pleasure that would be without thought, opinion, memory, or calculation "would not be that of a man, but of a mollusk or some other shell-fish like the oyster" (*zēn . . . ouk anthrōpou bion, alla tinos pleumonos*) (*Philebus* 21c), a life that would be so closed up in itself, so closed off to past and future pleasures, that it would end up experiencing no real pleasure at all. Socrates in the *Gorgias* argues in a similar way, contrasting, first, a temperate life, a temperate *bios*, with a licentious one, and then the *bios* or "life of philosophy" (*ton bion ton en philosophiai*) with the *bios* of pleasure and power, two competing lives that provoke the ultimate question, as Socrates phrases it to Callicles, of "how one ought to live [*pōs biōteon*]" (*Gorgias* 493e–494a, 492d, 500c–d).[5]

At issue in dialogues ranging from the *Apology* to the *Gorgias* to the *Laws* is the question of what makes life, *bios*, most *biōtos*, most livable or most worth living and, by contrast, what makes it least livable and least worthy of being lived. If, as Socrates says in the *Republic*, life becomes intolerable, unlivable, when that which constitutes the body's health has been destroyed, it becomes all the more unlivable or intolerable when that which constitutes the soul's health has been ruined (see *Republic* 407a, 445a–b, *Crito* 47d–e, *Symposium* 216a, *Letter VII* 340c). What makes life most livable (*biōtos*), most worth living, is, then, as Diotima says in the *Symposium*, the healthy soul in contemplation of the beautiful itself. It is "in that state of life [*entauta tou biou*] above all others," she says, that "a man finds it truly worthwhile to live [*bioton anthrōpon*], as he contemplates essential beauty" (*Symposium* 211d).[6]

A *bios* is, therefore, a life that is lived, for good or ill. There is the possibility of leading an examined life or an unexamined one, a life of wisdom or one of pleasure, a life of philosophy or one of pleasure and power, a just life or a tyrannical one, or, more specifically still, an "Orphic life" (in short, a vegetarian life) (*Laws* 782c), an Ionian as opposed to a Laconian life (*Laws* 680d), or, in the *Republic*, a "Homeric way of life [*hodon . . . biou*]" or a Pythagorean life (*Republic* 600a–b). There are, it seems, about as many *bioi* as there are distinct types of people, and these *bioi* can be compared and contrasted for the advantages or disadvantages, the virtues and vices, they entail. Determining which life, which *bios*, is best, which is most livable or tolerable, and which is worst and least livable, is a central concern of many Platonic dialogues, at once early, middle, and late. As Socrates says at the end of the *Republic*, the main concern of each of us, "neglecting all other studies, should [be] . . . to learn of and discover the man who will give him the ability and the knowledge to distinguish the life [*bios*] that is good from that which is bad" (*Republic* 618c; see 578c). In the *Laws*, often considered to be Plato's last dialogue, the Athenian says in a similar vein that the life that is "occupied with the care [*epimeleian*] of bodily and spiritual excellence in general" is the one "which most truly deserves the name 'life' [*bios eirēmenos orthotata*]" (*Laws* 807d)—the life, the *bios*, most worthy of the name, the life that is best for man to live, though not yet, as we will see, anything like what I will call later in this chapter *real life* or *life itself.*[7]

What is essential, then, as Socrates again says at the end of the *Republic*, is "to make a reasoned choice [*haireisthai*] between the better and the worse life [*bios*], with [one's] eyes fixed on the nature of [one's] soul, naming [*kalounta*] the worse life that which will tend to make it more unjust and the better that which will make it more just" (*Republic* 618c–e). For that is "the best choice," says Socrates, "both for life and death" (*hoti zōnti te kai teleutēsanti hautē kratistē hairesis*)—that is, "both in this life [*bios*], so far as may be, and in all the life [*bios*] to come" (*Republic* 618e–619a). In other words, it is important to know how to distinguish good lives, good *bioi*, from bad ones, holy lives from unholy ones, and then to live accordingly, because, Socrates is suggesting, our current *bios*, our "present life" (*ton nun bion*), determines to a large degree our future one(s) (*Republic* 534c). While it is a matter of much debate how we should understand Plato's various tales of reincarnation, how literally to take them, what is important here is the attribution of *multiple* lives, a plurality of *bioi*, to a single soul and the possibility of multiple lives or ways of life for a single soul. In the myth of Er at the end of the *Republic*, Plato describes the different possible fates for the soul in the next life as, precisely, *bioi*, as so many "patterns of lives" (*biōn paradeigmata*) laid out before the soul for its choosing (*Republic* 617d), as many lives, it seems, as there are "lots" (*klērous*) on the lap of Lachesis. There were, he says, "lives [*bioi*] of all kinds of animals [*zōa*] and all sorts of human lives" (*Republic* 618a), lives of swans, nightingales, and other musical animals, lives of lions, eagles, apes, and various kinds of wild animals (*Republic* 620a–d) and, if we wanted to augment the list with the help of the *Phaedo*, kites, ants, bees, and so on (*Phaedo* 81e–82b).[8] As for the human lives, or *bioi*, "there were lives of men of repute for their forms and beauty and bodily strength . . . and prowess and the high birth and the virtues of their ancestors, and others of ill repute in the same things, and similarly of women," lives of tyrants and lives of ordinary citizens who minded their own business (618a–b; see 620d).

There are thus many possible *bioi* for a soul to choose from in the afterlife, and a single soul can have many *bioi*, a first one as a human, a second or third as a beast, and vice versa. The *Phaedrus* speaks in a similar way of souls living a "first life" (*prōton bion*) and then, after being rewarded or punished for that life, being allowed "to draw lots and choose their second life [*tou deuterou biou*]," the result being that "a human soul may pass into the life of a beast [*eis thēriou bion*]" or, so long as the soul was once human, back again "from a beast into a man" (*Phaedrus* 249b; see *Timaeus* 42b). What is important here is that a soul appears to exist somewhat independently of its *bios* or its *bioi*, though it is each time marked or influenced, tainted or cleansed, and so punished or rewarded, it seems, by the kind of *bios* it lives. To each particular life there corresponds a kind of soul, a character of soul, but the soul can also change in character from one life to the next (*Laws* 803a). The possibility of variation and multiplicity is thus part and parcel of what it means to lead or live a *bios*, and this will be one of the key differences, as we will see, between *bios* and *zōē*. As Agamben rightly remarks, there is no plural of *zōē* in the Greek language, a little linguistic fact that corresponds, as we will see, to an important conceptual distinction.

But just so that this difference between *bios* and *zōē* does not become too neat or too rigid, there are, it has to be said, other meanings of *bios* that bring it much closer to *zōē*. In addition, then, to suggesting a way of life, *bios* sometimes means in the dialogues a livelihood, not one's way of living but one's "living" or means of living.[9] It also commonly denotes the span or duration of life, the time of a life, what we call a lifetime, a time that can be considered somewhat independently of the type of life that is lived (*Protagoras* 351d; *Symposium* 181d, 216e; *Republic* 367e, 576a–b; *Crito* 43b; *Euthydemus* 295a). As such, a *bios* can be measured, it can be longer or shorter; it lasts for a determinate length of time during which deeds are done and events suffered (see *Republic* 330e, 405b; *Phaedrus* 242a; *Phaedo* 114c); it is a span of time that typically ends in old age and, always, in death (*Republic* 330e, 469d; *Laws* 872e). Crito says to Socrates that because the ship has come in it appears that his *bios* will soon come to an end: "tomorrow," says Crito, "your life must end" (*ton bion se teleutan*) (*Crito* 43d). In the *Timaeus*, there is even the suggestion that not only every individual *zōon* but every species has a certain *bios*—that is, a certain lifespan— determined by its original constitution.[10]

To highlight this notion of duration, Plato often speaks of doing something or living in a certain way throughout one's lifetime or for one's whole life, *dia tou biou* or *dia pantos tou biou*. Crito credits Socrates with having "cared for virtue all his life [*dia pantos tou biou*]" (*Crito* 45d)—that is, having lived a virtuous life from the beginning to the end of his life.[11] Duration and the desire for continuity, in addition to multiplicity and the possibility of variation, are thus central as well to Plato's conception of *bios* as it is or as it should be lived, both for the individual and the city if not the human species as a whole.

Finally, it is because a *bios* has a specific length or time span that it can be seen, as it were, from afar, as a whole, with a beginning, a middle, and an end. In *Philebus*, Socrates says that pain is mixed with pleasure not only in tragedies and comedies on the stage but "in all the tragedy and comedy of life [*(en) tēi tou biou xympasēi tragōidiai kai kōmōidiai*]" (*Philebus* 50b). It is a similar emphasis on *bios* not as a way of life but as a length or span of time that allows the Athenian in the *Laws* to speak of old age as the sunset or the "evening of life" (*en dusmais tou biou*) (*Laws* 770a) and for him to suggest, not long thereafter, that because our *bioi* are so short, because we are so ephemeral as individual creatures, life, *bios*, must be passed on from one generation to the next.[12] The Athenian there speaks of the need for couples to leave their parents' homes and found their own "as if they had gone off to a colony," so that, "by begetting and rearing children," they might "hand on life, like a torch, from one generation to another [*kathaper lampada ton bion paradidontas allois ex allōn*]" (*Laws* 776b). Though not yet a principle of life—and note that Plato here uses *bios* rather than *zōē*—we seem to be getting closer to such a notion.

Bios/Bioō, Zōē/Zaō

While *bios* thus does indeed designate more often than not a certain way of life, while it does seem to be attached more often than not to a value judgment regarding *how* one should lead or live a life, there are passages where *zōē* is used in a similar way, places where it is almost a synonym for *bios*. In Book 7 of the *Republic*, for example, Socrates asks Glaucon about the "way of life [*bion*]" they have just laid out for the guardians—including ruling the city—and he suggests, in the same line, that only when such men are ruling will the city and its inhabitants achieve "a good and wise life" (*zōēs agathēs te kai emphronos*) (*Republic* 521a).[13] Plato obviously sees no problem using *bios* and *zōē* in the space of just five lines to say more or less the same thing.

As for the verbal analogues of *bios* and *zōē*, namely, *bioō* and *zaō*, there are even more overlaps and intersections, though that does not mean that there are not important differences between them.[14] As one might expect, *bioō* means, first of all, to live a life or a *bios*, one of those many *bioi* we just saw. One can thus speak of leading or living [*bioō*] one's life [*bios*] in a certain way—better or worse, in a more or less holy way, and so on. At the end of the *Euthyphro*, for example, Socrates feigns great disappointment at not having learned from Euthyphro the nature of the holy, for had he been so instructed he would have been able to reform his ways and, he says, "live a better life henceforth" (*kai ton allon bion hoti ameinon biōsoimēn*)—that is, live, *bioō*, a better *bios* (*Euthyphro* 16a).[15] The verb *bioō* can thus be used with *bios* as a direct object, or else without an object, in which case it is typically qualified with an adverb. In the *Phaedo*, Socrates thus speaks of what the afterlife holds in store for those "who have lived well and piously" (*hoi te kalōs kai hosiōs biōsantes*) and those who "have lived [*bebiōkenai*] neither well nor ill" (*Phaedo* 113d; see *Laws* 762e–763a, 865d–e).[16] Hence *bioō*, like *bios*, is usually identified with living in a particular way, living *well* or *ill*, not just simply living, even if there are places where the emphasis is placed on "living many years" (*ei polla etē bioiē*) or "living but a short time" (*oligon chronon biountōn*) (*Republic* 615c).

As for the verb *zaō*, or its infinitive *zēn*, it can be used much like *bioō*. We just saw how the verb *bioō* is often combined with *bios* to speak of the way one lives (*bioō*) a life (a *bios*). There are many other places, and often within the same dialogue, where it is the verb *zaō* that is combined with *bios* in a very similar way. Socrates asks Protarchus in the *Philebus* whether he would be willing to "live [*zēn*] [his] whole life [*ton bion hapanta*] in the enjoyment of the greatest pleasure" (*Philebus* 21a), while in the *Republic* he argues that the guardians, free of private possessions, "will live [*zēsousi*] a happier life [*biou*]" than the one that most people consider the happiest of all, namely, "the life [that] the victors at Olympia live [*hon (bion) hoi olympionikai zōsi*]" (*Republic* 465d; see *Republic* 406b, 495c; *Gorgias* 494c; *Phaedo* 95d; *Theaetetus* 177a; *Laws* 806d–e; *Letter VII* 327b).

Like *bioō*, *zaō* can also be used all by itself, or with a modifying adverb, to speak of living in a certain way, for example, "living well" or living "a good life" (*pros to eu zēn*) (*Republic* 387d), "living blissfully" (*eudaimonōs ezōn*) (*Letter VIII* 354d) or living "a life of pleasure" (*Phaedrus* 258e, see *Philebus* 47b), "living a life that is nobler and truer" (*kai zōien kallion kai orthoteron*) (*Menexenus* 248c–d), or else living as a parent (*Laws* 717b), or as a distinguished citizen (*Laws* 946e), or a man of virtue (*Menexenus* 236e), or, in Aristophanes's tale in the *Symposium*, fused together with another in a common life (*Symposium* 192e).[17] If, as we will see in a moment, a couple of passages in the dialogues suggest that *zaō* or *zēn* is indeed to be thought along the lines of bare life, there are many places where it is used to speak of a very particular kind of human life. In the *Timaeus*, Plato even speaks of living, *zaō*, "a most rational life" (*kata logon zōē*) (*Timaeus* 89d, *Laws* 689d), while in the *Laws* the Athenian speaks of "living" (*zēte*) one's life in accordance with law (*Laws* 771a). The verb *zaō* is thus hardly restricted to a pre- or apolitical context and is often used in contexts where it suggests anything but bare life.[18] The *Protagoras*, as if anticipating the *Statesman*, speaks of the city compelling young men once they have left school to "learn the laws, and to live [*zēn*] according to them as after a pattern [*kata toutous (tous nomous) zēin kata paradeigma*]" (*Protagoras* 326c–d; see also *Laws* 809d).[19]

We already saw that *bios* and *zōē* can sometimes be used more or less interchangeably, and sometimes in more or less the same breath; the same goes for *bioō* and *zaō*.[20] There are several places where the verb *bioō* appears to be more or less a synonym of *zaō*, places where differences in cadence or tense or else euphony, the presence, for example, of *bios* as a direct object that would resonate with *bioō*, no doubt played more of a role in Plato's word choice than any real difference in meaning between the two verbs.[21] In the *Protagoras*, Socrates goes back and forth between the two verbs to speak of "living well" (*eu zēin*) and having "lived well" (*eu . . . bebiōkenai*), having "lived pleasantly" (*hēdeōs bious*) and having "lived [*zōē*] in distress and anguish [*aniōmenos te kai odunōmenos*]" (*Protagoras* 351b). In the *Laches*, Nicias says that the person who converses with Socrates is "led into giving an account of himself, of the manner in which he now spends his days [*zēi*], and of the kind of life he has lived [*bion bebiōken*] hitherto" (*Laches* 187e–188a). A single speaker can thus shift from one word to the other even in the course of a single sentence.[22] To give just one final example, in Socrates's response to Thrasymachus's tirade at the end of Book 1 of the *Republic*, we find, in the space of just a few lines, not only the two nouns, *bios* and *zōē*, used in a quasi-synonymous way, but the two verbs, *zaō* and *bioō*—as well as the noun *diagōgē*, which is also used in Plato to suggest the course or conduct of one's life. Socrates says to Thrasymachus: "Do you think it is a small matter that you are attempting to determine and not the entire conduct of life [*biou diagōgēn*] that for each of us would make living life [*diagomenos . . . zōēn zōiē*] most worthwhile? . . . You seem to . . . care nothing for us and so feel no concern whether we are going to live [*biōsometha*] worse or better lives in our ignorance of what you affirm that you know" (*Republic* 344e).

Zōē *as Bare Life*

I will not multiply the examples any further because the point should be pretty clear: *bios* and *zōē*, *bioō* and *zaō*, overlap in many different ways and places. The distinctions often seem to be dictated more by grammar, as I have suggested, than anything else. But they also vary in frequency and according to dialogue in a way that seems to suggest something else. Interestingly, *zōē* is used much less frequently in Plato, only eighteen times, as opposed to *bios*, which, as I mentioned earlier, is used eighty-five times, more than four times more than *zōē*. But even more telling is the fact that *zōē* is used almost exclusively in what are considered to be late middle or late dialogues—that is, in the cluster of seemingly related dialogues that have been at the center of this work, *Phaedrus* (1), *Phaedo* (4), *Republic* (2), *Laws* (5), *Sophist* (1), *Timaeus* (2), and, if one wants to count it, *Epinomis* (2). The one apparent exception to this rule would be a single use in *Alcibiades I*—a little lexical fact that may lend some credence to the hypothesis of some scholars, including Foucault, that this was an early dialogue that was *rewritten* by Plato around the time of *Laws*.²³ While *zōē* and *bios* and their corresponding verbal forms thus often overlap, there are some significant distinctions that, while never hard and fast, can help us understand this almost exclusive restriction of the noun *zōē* to later dialogues, where—and this will be my hypothesis—we begin to see a shift in the word's meaning, a transformation from what might be called bare life to *real life*, a transformation that will be important for the neo-Platonic tradition, not to mention the entire Christian tradition that comes out of it or makes reference to it.

Here, then, are just five places where it is tempting to see something like a notion of life as bare life coming into relief, five places where *zōē* or *zaō* seems to describe more an activity or a process of living than *bios* or *bioō* do, an activity or a process that can begin and end but typically does not have a span or duration as *bios* does, an activity or a state of being alive rather than a type of life or a particular way of living over a period of time. For example, in the *Crito*, the Laws of Athens say to Socrates in his prosopopeia of them: "Ah, Socrates, be guided by us who tended your infancy. Care neither for your children nor for life [*mēte to zēn*] nor for anything else more than for the right that when you come to the home of the dead, you may have all these things to say in your own defense" (*Crito* 54b; see *Gorgias* 512a–b). In this context, the Laws seem to be cautioning Socrates not to cling to life for the sake of life. *To zēn*, living, here seems to mean something like *mere living*, living for the sake of living, which is why, in the rest of the dialogue, *zēn* is distinguished from *eu zēn*—that is, from living well, as if it takes the qualifier *eu* to turn mere living into a genuinely human life or a life worthy of the name. As Socrates says in his response to Crito's plea to escape from prison, "it is not living [*to zēn*], but living well [*eu zēn*]," by which he means living justly, that is most important (*Crito* 48b), a distinction that Aristotle will himself use through-out the *Politics* (see, for example, 1252b30 and 1257b10). Later in the dialogue, these same Laws of Athens use both *bios* and *zēn* in a single sentence, providing a nice contrast between the two. They say to Socrates: "But will no one say that you, an old man, who had probably a short time yet to live [that is, whose *bios* had so little time left] [*smikrou chronou tōi biōi*], clung to life [*epithumein zēn*] with such shameless greed that you transgressed the highest laws?" (*Crito* 53d–e).²⁴ In order to continue to lead an honorable *bios* or a *bios* of virtue for the time he has left, Socrates must not, the Laws are suggesting, cling to life or desire to live, *zēn*, for the mere sake of living.[25]

Second example: Whenever Plato wants to talk of a historical figure who is still living, or else no longer living, his verb of choice is invariably *zō* rather than *bioō*, as if this former were better suited to speak of the mere fact of living or being alive. Hence, Socrates in the *Theaetetus* says that Protagoras would have defended himself against their arguments if he "were alive [*ezē*]" (*Theaetetus* 164e), while in *Republic* he asks whether Homer was ever credited, "while he lived [*zōn*]," with educating others and passing on a Homeric way of life (*Republic* 600 a–b). In a similar vein, some form of *zaō*, rather than *bioō*, is used to characterize living as a captive (*Republic* 468a, *Seventh Letter* 347e) or, indeed, as a survivor, someone who is still living despite some danger, threat, or trial (*Laws* 958e).[26]

Third example: In a passage from the *Gorgias*, Callicles seems to evoke the paradox, at least for him, of *living*—simply living—within the state without power or possessions. He suggests that a man would be a fool to live in a way that makes him vulnerable to being "stripped by his enemies of all his substance [*ousian*], and to live in his city [*zēin en tēi polei*] as an absolute outcast" (*Gorgias* 486b–c). The fact of living in the city without power or property or substance, as a man without qualities, as it were, is here emphasized, it seems, by the use of *zēn* without any qualifier, *zaō* as opposed to *bioō*.

Fourth: It is perhaps not insignificant that, in the *Laws*, the life that is lived in the golden Age of Kronos is described not as a *bios* but as a *zōē*. The Athenian there speaks of tradition telling us "how blissful was the life of men in that age [*tēs tōn tote makarias zōēs*]" (*Laws* 713c). A life before or without the *polis*, before or without the arts, a life where everything is furnished automatically or spontaneously to mankind, is thus, here, a *zōē*—though, it has to be noted, a similar passage in the *Statesman* speaks, as we saw in Chapter 2, of the "spontaneous *bios*" (*automatou peri biou*) of mankind (*Statesman* 271e).

Fifth and finally: The verb *zaō*, which, as I said, seems to denote more of a process or a state, which seems to be, so to speak, *more verbal* than *bioō*, is contrasted in the *Laws* with a life, a *bios*, that comes to an end. The Athenian there speaks of the danger of honoring "with hymns and praises those still living [*eti zōntas*], before they have traversed the whole of life [*hapanta . . . ton bion diadramōn*] and reached a noble end [*telos epistēsetai kalon*]" (*Laws* 802a). In other words, it is dangerous to praise those still living, *zōntas*, before they have reached the end of their *bios*.[27]

But this final passage perhaps tells us something more. Because, as I have argued, *zaō* and *zōē* both suggest, much more than *bios* and *bioō*, a process or activity or state of living, then it would seem that *zaō*, rather than *bioō*, would be the real opposite of being dead, while *zōē*, rather than *bios*, would be the opposite of death or *thanatos*. And that is exactly what we find throughout the dialogues. In the *Republic*, guardians must receive honors and rewards "in life and in death" (*kai zōnti kai teleutēsanti*)—that is, while living and after they have died (*Republic* 414a)—a phrase that is repeated in almost identical form throughout the dialogues.[28] Sometimes, the act of *dying* is contrasted with *living* as *bioō*, as when Socrates says near the end of the *Apology*, "I go to die [*apothanoumenōi*], and you to live [*hēmin de biōsomenois*]" (*Apology* 42a), but when it comes to *being dead* and *being alive* Plato's verbs of predilection are invariably *teleutaō* and *zaō* (not *bioō*)—that is, having reached an end, come to fulfillment, in short, died, as opposed to continuing to live.[29]

Living, *zaō*, is thus the opposite of being dead and life, *zōē*, the opposite of *thanatos*. Socrates asks in the *Phaedo*, "Now do you tell me in this way about life and death [*peri zōēs kai thanatou*]. Do you not say that living is the opposite of being dead [*ouk enantion men phēis tōi zēn to tethnanai einai*]?" (*Phaedo* 71d). Socrates asks even more succinctly just a few pages later: "Is there anything that is the opposite of life [*poteron d'esti ti zōēi enantion, ē ouden*]?" and the answer, again, is *thanatos* (*Phaedo* 105d).[30]

Life, *zōē*, would thus be the opposite of death, *thanatos*, and living, *zaō* or *zēn*, the opposite of being dead. That is, as I have suggested, because *zōē* and *zaō* describe an activity or function of the soul in a way neither *bios* nor *bioō* do. In the *Republic*, Socrates will thus speculate that life or living (*to zēn*) is the *ergon*, the very work or function—or at least one of the works or functions, along with management, rule, and deliberation—of the soul (*Republic* 353d). Living is that which the soul does or brings about. The soul is what makes a body live, an affirmation that is repeated in the *Phaedo*: "What causes the body in which it is to be alive [*hōi an ti engenētai sōmati zōn estai*]?" asks Socrates, and the answer is, of course, the soul. For "if the soul takes possession of anything [*hē psychē ara hō ti an autē kataschēi*], it always brings it to life [*aei hēkei ep' ekeino pherousa zōēn*]" (*Phaedo* 105d). A *zōon*, then, is the result of the soul taking control of the body in this way. As we read in the *Timaeus*, it is "that compound of soul and body which we call the 'living creature' [*zōon ho kaloumen*]" (*Timaeus* 87e). In the *Cratylus*, this taking possession of the body by the soul makes the body not only live but move: "Do you think there is anything which holds and carries the whole nature of the body, so that it lives and moves [*zēn kai periienai*], except the soul?" (*Cratylus* 400a)[31] Despite the many places of cross-over or convergence, there does seem to be a difference between *bios* and *bioō*, on the one hand, and *zōē* and *zaō*, on the other.

Zōē *as Life Itself*

Zōē, rather than *bios*, then, is the work or the activity of the soul; it is what comes about because of the soul. For the moment, however, it is only the *conjunction* of body and soul that allows one to say that something is alive. We are, in some sense, back to the passage in the *Sophist* with which I began this entire work, the one in which a *zōon* was used as an example of the conjunction of a body in the visible realm and a soul in the invisible one. One could put it like this: A soul—that is, a *psychē*—takes hold of a body, a *sōma*, and makes it live, gives it life (*zōē*), and this composite of body and soul, this *zōon*, will then necessarily have a *bios*, a *bios* with a certain moral character and a certain duration, indeed, as we have seen, a *bios* that can change from one life to the next. But the question then becomes just exactly what status the soul has independent of these bodies and so independent of all these particular *bioi*. Can it be said to *be*, all on its own, apart from all bodies, and, especially, can it ever be said to *live* apart from all the bodies to which it gives life?

In the *Phaedo*, Socrates begins to provide an answer to this question in one of his first proofs for the immortality of the soul. Though the argument will be insufficient for proving the soul's immortality and it is doubtful whether Socrates really even proposes it in all seriousness, the terms and categories he uses are nonetheless significant. Employing the words for being dead and living that we have just seen, Socrates argues that insofar as opposites are generated from opposites, life must be generated from death and death from life. It is "from the dead," he says, that "living things and persons are generated" (*ek tōn tethneōtōn ara . . . ta zōnta te kai hoi zōntes gignontai*) (*Phaedo* 71d) and it is this process of being generated from the dead that can be called "coming to life again" (*anabiōskesthai*)—a form of *bioō* and not *zaō*, interestingly, because what we are talking about here is not just living again but coming to have a new life, a new *bios* (*Phaedo* 71e).[32] Now, if all this is true, says Socrates, then "the return to life [*anabiōskesthai*] is an actual fact, and it is a fact that the living are generated from the dead and that the souls of the dead exist [*tas tōn tethneōtōn psychas einai*]" (*Phaedo* 72d). As he will put it later, if "every living being [*pan to zōn*] is born from the dead," then "the soul exists [*estin*] before birth, and, when it comes into life [*eis to zēn*] and is born"—the verb *zēn* being used here, not *bioō*—then "it cannot be born from anything else than death [*thanatou*] and a state of death" and so "must also exist [*einai*] after dying, since it must be born again" (*Phaedo* 77d).[33]

The soul thus *exists*, it *is*, both before the birth and after the death of the *zōon* it causes to live. Living and being thus appear, at least for the moment, to be distinct.[34] When the soul is separated from the body, neither it nor the body can be said to live, though both can be said to *be*. Neither has a *bios*, of course, and neither can be said to have or to participate in *zōē*. Life, here, would seem to be restricted to what we *call* life, to the time when a soul takes possession of a body to make it live.

But one can already anticipate the disquiet or the concern, perhaps even the anxiety. Were Plato to have left things like this, then the whole category of life, whether *bios* or *zōē*, would have been restricted to the time between birth and death, to the time when a *zōon* is, as we say, living, alive rather than dead. The soul in the time before birth and after death would then *be*, but it would not be *alive*. This would be problematic enough for a philosophy that identifies human life with the cultivation of virtue attached to an invisible, immortal soul; it would be downright deadly, however, for a philosophy that identifies *man himself, anthrōpos* itself, with that soul. Within life or within a lifetime (within a *bios*) it may be *zōē*, the result of the soul's activity, that defines one as a *zōon*, as a living being and not a lifeless body, but it is ultimately the soul, the immortal soul, that defines one as an *anthrōpos*—that is, as a human being. The immortal soul takes hold of the body and makes it *live*, gives it life, *zōē*, and provides it with a life, a *bios*, but it is this same soul that makes the human what it essentially *is*, quite apart from its body, its *zōē*, or its *bios*. As the Athenian says in the *Laws*,

> we must believe [the lawgiver] when he asserts that the soul is wholly superior to the body, and that in actual life [*en autōi te tōi biōi*] what makes each of us to be what he is is nothing else than the soul, while the body is a semblance [*indallomenon*] which attends on each of us, it being well said that the bodily corpses are images [*eidōla*] of the dead, but that which is the real self of each of us, and which we term the immortal soul [*athanaton einai psychēn eponomazomenon*], departs to the presence of other gods. (*Laws* 959c)

Put more succinctly in *Alcibiades I*—yet again, notice, *Alcibiades I*—man is "nothing else than soul" (*Alcibiades I* 130c).[35]

If man *is* his soul, then the utmost care must be taken for that soul, both for the sake of this *bios* and the *bioi* to come—that is, for the sake of what we continue to *call* life. As Socrates argues, once again in the *Phaedo*, "if the soul is immortal [*hē psychē athanatos*], we must care for it, not only in respect to this time, which we call life [*oukh hyper tou chronou toutou monon en hōi kaloumen to zēn*], but in respect to all time [*all' hyper tou pantos*], and if we neglect it, the danger now appears to be terrible" (*Phaedo* 107c). There is thus the time we *call* life and there is the time—all of time—in which the soul, independently of its lives, simply *is* or *exists*. The immortality of the soul means that the soul is or exists not only during the time of life but before and after, in the soul's pre-life and afterlife. But there is also the implication here—or at least the temptation to think—that there is, in addition to what is *called* life, in addition to this *so-called* life, another life, a real life or *life itself*, a genuine life in the life of the soul. Though this is only implied or suggested in this passage, it is made explicit elsewhere. In the *Republic*, for example, it is said that only the one whose soul comes to dwell with the things to which it is akin—that is, intelligible things—will "beget intelligence and truth, attain to knowledge and truly live [*gnoiē te kai alēthōs zōiē*] and grow and so find surcease from his travail of soul" (*Republic* 490b).

Alēthōs zōiē: The phrase suggests not just a way of life within this life but a different kind of living. Only the one whose soul comes to know the truth can truly live. It seems hardly a coincidence, then, that, in the *Phaedo*, the dialogue in which the question of the existence of the soul before birth and after death must be confronted most directly, Socrates would advance the thought that the soul itself not only *is* in the time before birth or after death but actually *lives*. The fact that Socrates is speaking here in a myth and that he makes the attribution just in passing does not diminish in the least the significance of the terms he is using or the claim he is making:

> But those who are found to have excelled in holy living [*hosiōs biōnai*] are freed from these regions within the earth and are released as from prisons; they mount upward into their pure abode and dwell upon the earth. And of these, all who have duly purified themselves by philosophy live henceforth altogether without bodies [*aneu te sōmatōn zōsi*], and pass to still more beautiful abodes which it is not easy to describe. (*Phaedo* 114b–c)

Here we are, finally, with a notion of life or of living detached from bodies, a notion of *zōē* attached only to the soul. The soul *is* before and after it makes a body live, but now it seems that, in between the various lives it animates, in between those *bioi* where the body has *zōē*, it itself also *lives*. Previously, it was only when the soul took hold of a body that we could talk about life, about something being alive, something being a *zōon*. Now, finally, we can we talk about the soul itself as *alive*.

One might be tempted, at this point, to exclude this rather singular use of *zōē* or *zēn* from the rest of the dialogues, and maybe even the entire Pythagorean-inspired *Phaedo* from the Platonic corpus. But one would then be faced with having to exclude as well this passage from *Phaedrus* where Socrates seems to give to the soul not only immortality but self-movement and, finally, life:

> Every soul is immortal [*athanatos*]. For that which is ever moving [*aeikinēton*] is immortal; but that which moves something else or is moved by something else, when it ceases to move, ceases to live [*paulon echei zōēs*: ceases to have life, *zōē*]. Only that which moves itself, since it does not leave itself, never ceases to move, [and, by implication, never ceases to live] and this is also the source and beginning of motion for all other things which have motion. (*Phaedrus* 245c)[36]

Insofar as it is the soul, not the body, as we saw, that makes things come alive, that allows one to *call* something *alive*—that is, that allows for something to have *zōē* and to live a *bios*—it is not surprising to see Plato wanting to attribute not just *being* but *life* to the soul, a life before what we call life. But notice what has now happened to this notion of life or *zōē* in Plato's own internal *gigantomachia*. By distinguishing what is visible and mortal from what is invisible and immortal, and by then attributing a kind of life to this latter, two types of life, two kinds of *zōē*—the same *zōē*, recall, that has no plural in Greek—begin to appear, the one very different from the other: There would be so-called life, the life of this life, not exactly bare or mere life but the semblance or simulacrum of life, and then the life of the soul—that is, a superior and independent life, in other words, *real life* or *life itself*.[37]

One might, of course, claim that this "life" of the soul is but a metaphor, a live metaphor for life but a metaphor nonetheless. But one would then have to explain how it is that this so-called life, this literal life, we might say, gets its sense or its meaning from the metaphorical life that

comes both before and after it, how it is that the life that is on the side of
the metaphorical comes to have so much of the value and meaning of life—
motion, reality, truth—while that which is on the literal side of life has so
little. And, in addition to all that, one would have to explain how it is that
both of these lives seem to get their meaning from the *form* of life itself.
For, in the *Phaedo* again, it appears that there is not only the life we call
life, and not only the life of the soul, but a form of life, an *eidos* of *zōē*, that
would give meaning (and life) to both kinds of life—that is, to everything
that is *called* life and everything that life truly *is*. Socrates says:

> All, I think, would agree that God and the principle of life [*auto to tēs*
> *zōēs eidos*], and anything else that is immortal [*athanaton*], can never
> perish [*apullusthai*]. . . . Since, then, the immortal is also indestructible
> [*adiaphthoran*], would not the soul, if it is immortal, be also imperish-
> able [*anōlethros*]. (*Phaedo* 106d)

There is thus an *eidos* of *zōē*, something for which there is and really could
be no equivalent with regard to *bios*. It was, it seems, a thinking of *zōē* as
mere or bare life, bare life as opposed to a properly human life, a singular,
invariable *zōē* as distinguished from a multiple and varied *bios*, that made
possible the transformation of *zōē* into something that goes beyond mere
human life, a transformation into, first, a life of the soul and then life in
the form of a form or an *eidos*, an *eidos* of life, in which all *zōa*, all living
things, all bodies *and* souls, would participate. As we saw back in Chapter 2
when we were trying to understand what Plato means by the term *zōon*, it
is the participation in life, in the form of life, that allows one to *call* something alive—that is, a *zōon*: "For everything . . . which partakes of life may
justly and with perfect truth be termed a living creature" (*pan . . . ho ti per*
an metaschēi tou zēn, zōon . . . an en dikēi legoito orthotata) (*Timaeus* 77b).[38]
There is what is called life, there is life beyond this so-called life, and
then there is a form of life, the Life Form, that gives meaning and, it seems,
life to all these lives.

The Life of Being

As Socrates claims in the *Phaedo*, God is immortal, as is the *eidos* of *zōē*,
the principle or form of *zōē*; neither can be destroyed, and so both are and
will continue to be. Both will continue to participate in being. The great
outstanding question is now whether *being itself* has life—not just that be-

ing that is the soul but being itself and all the other forms that are. The question is whether being itself is alive, whether it can be said to live. Well, that is precisely one of the questions asked by the Stranger in the *Sophist*, the dialogue that not only names but, we can now see, is central to the gigantomachia over being and life that Derrida identified in *H. C. for Life*. The Stranger there exclaims:

> But for heaven's sake, shall we let ourselves easily be persuaded that motion [*kinēsin*] and life [*zōēn*] and soul and mind are really not present to absolute being [*tōi pantelōs onti*], that it neither lives [*zēn*] nor thinks [*phronein*], but awful and holy, devoid of mind, is fixed and immovable? . . . Shall we say that it has mind, but not life [*zōēn*]? (*Sophist* 248e–249a)

Plato is thus ultimately tempted, it seems, by the thought that the soul itself, apart from the body, and that *being itself*, apart from everything that *is*, may be alive. But that would then mean that there really are two kinds of life, that is, what we *call* life, a life that always mingles with death, a life on the side of becoming, and *life itself*, the *eidos* of life, life on the side of being, a life totally detached from becoming and from death. There would be life, then, what we *call* life, and then *life itself*.

Without having to affirm some crude two world Platonic theory, a world of Forms and a world of particulars, this analysis of *bios* and *zōē* and their verbal counterparts does seem to suggest a difference between two forms of life, between, on the one hand, a sort of bare life to which we must not cling or else a human life that we must cultivate and, on the other, *life itself*—that is, real life, which, when pursued as far as possible in *this* life, through philosophy, for example, is the only thing that makes life truly worthy of the name and truly *worth living*. The unexamined life is not worth living, to be sure, but then the examined life is perhaps worth living only to the extent that it imitates life itself, that life that is most worth living even though it is a life that can never truly be lived.

As I come to my conclusion, let me simply suggest that all this might help explain why Plato in the *Timaeus*, one of the dialogues that Plotinus refers to most, would attribute life not just to *zōa*, to creatures, and not just to that mega-*zōon* that is the cosmos, but to the intelligible model of which that cosmos is an imitation. For that model is said by Timaeus to be an "all-perfect Living Creature [*tōi pantelei zōōi*]" (*Timaeus* 31a–b) that "embraces and contains within itself all the intelligible Living Creatures [*ta* . . .

noēta zōa panta ekeino en heautōi perilabon echei], just as the Universe contains us and all other visible creatures that have been fashioned [*kathaper hode ho kosmos hēmas hōsa te alla thremmata xynestēken horata*]" (*Timaeus* 30d). There are thus *two* living creatures, one within time, we might say, and one outside of time. The first of these two creatures, the one that is in time, the cosmos, can be *called* alive because it was made in imitation of the one that is not in time—that is, the one that functions as a model for all those things that are said to live, the one, the only one, then, that truly is and truly lives as the essence or source of all living things, the one that contains not all *thnēta zōa*, all mortal creatures, the products of soul and body, but all *noēta zōa*, all *intelligible living creatures*. Things live, *zōa* are, therefore, only to the extent that they are modeled upon "an eternal Living Creature [*zōon aidion*]," that is, only insofar as they are modeled on a Living Creature that does all its living *outside of time*. As Timaeus explains:

> And when the Father that engendered it [that is, the cosmos] perceived it in motion [*kinēthen*] and alive [*zōn*], a thing of joy to the eternal gods [*aidiōn theōn*], He too rejoiced; and being well-pleased He designed to make it resemble its Model still more closely. Accordingly, seeing that that Model is an eternal Living Creature [*zōon aidion*], He set about making this Universe, as far as He could, of a like kind. But inasmuch as the nature of the Living Creature [*zōou*] was eternal [*aiōnios*], this quality it was impossible to attach in its entirety to what is generated; wherefore He planned to make a movable image of Eternity, and, as He set in order the Heaven, of that Eternity [*aiōnos*] which abides in unity He had an eternal image [*aiōnion eikona*], moving according to number, even that which we have named Time. (*Timaeus* 37d–e)[39]

The universe—a *zōon*—is thus generated on the basis of the ungenerated, which alone is eternal, and the time of the universe, our time, the time of Zeus, we might say, is created as or in the image of what is eternal and living, truly living, like the time of Kronos. Life, what we call life, is thus modeled on *real life*, on a *zōē* that is not riveted—as bare life would be—to mere existence, mere living, but is, on the contrary, completely separated from all bodies, a *zōē* that is not bare life, then, but *life itself*.

Time is thus the time of what we *call* life, while life in time is modeled on that true life that is outside of time and outside our lives. That is just one way, I think, that Plato can be said to have initiated the great gigantomachia that Derrida speaks of in *H. C. for Life*, the great battle to know whether "it is necessary to think being [*l'être*] before life, beings [*l'étant*] before the living, or the reverse." It is beyond the scope of this work to look at how that gigantomachia continues in someone like Plotinus, who is a close reader of these lines from the *Timaeus* though also, of course, of so much of Plato's work. Let me simply conclude with a single quote where some of the stakes of what I have been arguing come into relief in a rather striking way. In what is thought to be a late treatise of Plotinus's, positioned by Porphyry as Book 4 of the First Ennead, Plotinus says the following:

> The term "life" is used in many different senses, distinguished according to the rank of the things to which it is applied, first, second and so on; and "living" means different things in different contexts; it is used in one way of plants, in another of irrational animals, in various ways of things distinguished from each other by the clarity or dimness of their life . . . We have often said that the perfect life [*teleia zōē*], the true [*alēthinē*], real [*ontōs*] life, is in that transcendent [*ekeinēi*] intelligible reality, and that other lives are incomplete, traces of life [*indalēmata zōēs*], not perfect or pure and no more life than its opposite. (I.4.3.19–37)

There is here, then, as in Plato, real life and so-called life, perfect life and imperfect life, a life that is transcendent in its intelligible reality and then the traces—the *indalēmata*, the same word that was translated at *Laws* 959c as "semblances"—of that first and most perfect life. *Indalēmata* as either traces of life *or* mere semblances of life, hypostases of life *or* imitations or simulacra of it: everything is at stake in this simple translation, everything from the relationship between the two kinds of life we have been following throughout this chapter to the relationship, perhaps, between the Age of Kronos and the Age of Zeus. *Indalēmata* as either traces *or* semblances, *ichnē* or *eidōla*, a thinking of Life imitating "life," a Life that gets its name and its meaning from this "life" that is lived, or a "life" that imitates Life, a "life" that is called life only to the extent that it imitates that true Life that is never fully lived. It's a difference that alone is sufficient to give renewed energy and life to the gigantomachia.

Conclusion: Life on the Line

> An irreducible rupture and excess can always be produced within an
> era, at a certain point of its text (for example, in the "Platonic" fabric
> of "Plotinism"). Already in Plato's text, no doubt.
>
> —JACQUES DERRIDA, "Form and Meaning," in *Margins of Philosophy*

> Always prefer life [*la vie*] and constantly affirm survival [*la survie*] . . .
>
> —JACQUES DERRIDA, *Final Words*[1]

There are, as we have seen, many life forms in Plato. There is, first, what
is *called* life, either zōē or *bios*, though most often *bios*, a life that can be either
long or short, well or poorly lived, a life of pleasure or wisdom, the life of
a tyrant or a philosopher. Then there is, in addition to these different ways
of life, the mere fact or the simple activity of living, something resembling
what Agamben calls *bare life*, a life that comes before and is indifferent to
the kind of life that is led. But then there is also—and this is truly unique
to Plato—what was called in the previous chapter *real life* or *life itself*, a life
over and above everything we call life, a life independent of and transcendent to that life, a life that provides the measure for and gives meaning to
all other life forms. It is this particular life form, this Form of life, as we
saw, that seems to become detached in Plato, and particularly in later dialogues, from everything called life, even if everything called life seems to
get its meaning and, in some sense, even its name, from that real life that
would have preceded it.

Real life or life itself gives meaning to everything called life: This is, to be sure, a rather Platonist reading or understanding of the dialogues. It is also, as I hope the previous analyses have demonstrated, one that is difficult to deny or to refute. But the fact that all these different kinds or forms of life are called or named, precisely, *life*, suggests that another reading is possible, one in which what is derived might be taken for the origin and the origin for what is derived. Instead of reading real life or life itself as that which precedes and gives meaning to what is called life, instead of understanding so-called life as the mere supplement to life itself, the fallen version of it, we might think of life itself as the "invention" of what is called life, the hypostasized version of this life, an empty concept or verbal construct that has nonetheless completely transformed the Western philosophical and theological tradition. On this reading, *life itself* is the least real and the most untrue life, a mere fiction or phantasm, an illusion, in short, of what is called life. It is a reading or an interpretation that might be identified as the critique or the reversal of Platonism, the revenge of the Giants against the Olympians, as it were: to these latter who would claim that the real world and real life are elsewhere, the former would argue that that "elsewhere" is simply an invention, a figment of the Olympian imagination.

But what is it, we might ask, that draws the lines between these various life forms and between all the concepts or distinctions upon which they rely (real/fictive, discovered/invented, life/non-life, and so on)? If a certain thinking of life seems to subtend all these distinctions, then it is perhaps possible to think that the differences between these various life forms are themselves the effect of another life, a life that has been forgotten or concealed or overlooked precisely because it is everywhere, another "kind" of life that never appears as such in the dialogues, indeed that cannot appear as such but that is absolutely essential to the dialogues, a life, in short, that loves to hide in the distinctions it makes and the lines it draws. Because one of the lines this other life draws would be the one between life and death, a line that we have been following throughout this entire work, then this *other life* would be unthinkable outside or without a relationship to death. This would thus be a life that always negotiates with death and with everything associated with it (absence, withdrawal, artifice, writing, and so on), a life that is, therefore, always and from the beginning compromised by death, a life that must always be forgotten in order for *life itself* to come on the scene, a life that thus *draws the line* between itself and all these other kinds of lives, a life, in short, that does not exist beyond all the lines it draws. This life would not *be*, then, it would not be another life form or Form of life, even though it would give rise to all these forms. It would be neither being nor a being and yet it would be the most "life-giving" of all these lives since it gives or draws— at the same time as it undoes or withdraws from—all these lives and all the distinctions between them. It is a life that weaves and unweaves, as it were, all these lives and all these lines. This other life, as a life that loves to hide, would be the life that draws and redraws the line between all these different kinds of lives as well as between life and death, the human and the animal, life and non-life, the organic and the inorganic, *physis* and *technē*, being and logos, and so on. As Derrida writes in a very late text, *Rogues*:

> What applies here to *physis*, to *phuein*, applies also to life, understood
> before any opposition between life (*bios* or *zōē*) and its others
> (spirit, culture, the symbolic, the specter, or death). In this sense, if
> auto-immunity is *physio*-logical, *bio*-logical, or *zoo*-logical, it precedes
> or anticipates all these oppositions. My questions concerning "political"
> auto-immunity thus concerned precisely the relationship between the
> *politikon*, *physis*, and *bios* or *zōē*, life-death.[2]

In the introduction to this work, I identified this other life with what Derrida called in a seminar of 1975–76, and then again here in *Rogues*, *life death*. This is a thinking of life that Derrida pursued not only in that seminar but, truth be told, in all his works, from as early as *Of Grammatology* (1967) right up to *Rogues* (2003) and his very last seminars. In the first of these works, for example, Derrida seems to endorse the view, which he arrives at through a reading of Rousseau, that "the principle of life is . . . confounded with the principle of death," which means that "death is not the simple outside of life," that "death is the movement of difference to the extent that that movement is necessarily finite," but, also, that death, by writing, "inaugurates life."[3] Hence a certain thinking of death or *life death* inaugurates or opens up life at the same time as it compromises it and marks it from the start with finitude. *Life death* would thus name a life that is inseparable from death, a life that, through this inaugural gesture, opens time without being outside it or transcendent to it.

Some thirty-five years later, in *Rogues*, Derrida would still be turning around this same principle of *life death*, this time under the name of *autoimmunity*, a thinking of life as what is, in principle and not as a matter of accident, confounded with death, a thinking of life that compromises or undoes itself from the inside, as it were, and from the origin. This concept or quasi-concept of *life death* would be the final "form" of life to be understood or glimpsed in Plato, even if it is not another Platonic form. Instead of life itself, then, instead of the Form of life, there would be *life death* as what exceeds the neat boundaries between life and death, a life before all life forms or ideas, in short, a life before conception.

Because a reading of Derrida has been central to this work, as I suggested in the introduction and as the reader will have seen evidence of throughout, it is perhaps worth taking a moment to underscore how this thinking of life was central to Derrida's work—including his seminars—right up until the end. Throughout his very final seminar, *The Beast and the Sovereign*, for example, Derrida questions the putatively single and indivisible line that is commonly drawn by philosophers between the human and the animal (in general). Derrida thus questions the rigor and legitimacy of all those discourses that tend to draw the line between humans and the animal on the basis of attributes or possessions such as language, technology, mourning, tears, law, a relation to death, or, indeed, though this is often less obvious, life—that is, real life or life itself. Though the term *life* usually does not appear in the long list of attributes that Derrida claims philosophy grants the human and denies the animal, it is implicit in all of his analyses. From this perspective, *real life* or *life itself* would be what is proper to man, a thinking that we saw to be central to Plato. Other living things, other animals, would be alive, of course, but they would not have access to the kind of real life that man does. That question of the boundary between the human and the animal has, obviously, haunted this entire work as well.

This emphasis on life also underscores just how Derrida's final seminar, *The Beast and the Sovereign*, arose out of his preceding seminar on *The Death Penalty*, where, as Derrida demonstrates, the most rigorous proponents of the death penalty, and Kant in particular, claimed that it is only by being willing to give up one's life to the state that the citizen can go beyond mere biological, animal, or phenomenal life, everything, in short, that is *called* life, in order to accede to freedom and ethical life, to a value beyond life, to something like life itself. The death penalty would thus be, accordingly, what is proper to man, since it is founded upon a sacrificial logic that elevates man above the animal and above animal life by giving him access to law and to life in the proper sense of these terms.[4]

It would not be an exaggeration to say that deconstruction itself, Derrida's entire critique of the so-called metaphysics of presence, was from the beginning a deconstruction of the concept of life, a critique of life's supposed purity and difference from death, its exclusion of death and everything associated with it (beginning with absence and alterity, difference and writing). Without being a vitalist thinker—indeed his thinking runs to the contrary—Derrida was, from start to finish, a thinker of life or, rather, of *life death* as that which compromises from the start anything like a pure life or a purely living present. He was a thinker who tried to find a certain articulation or inscription of *life death* in various thinkers throughout the Western philosophical tradition, beginning with Plato, and who then tried to develop an explicit account of what was usually not explicit in these thinkers, an explicit account of what was not explicit and an account of why it could not be explicit in these thinkers.

If I have been able to show in this work that all these other life forms really do have a place in Plato, including—the most Platonic of them all—*life itself*, I will not be able to show with the same degree of clarity this other kind of life that I have called, following Derrida, *life death*. That is because this other kind of life loves to hide, as I said, but also because Plato himself nowhere makes explicit this other kind of life, never makes an explicit argument for it and never refers to it as such. He never speaks of life as he speaks of *khōra*, for example, as "what must be called always by the same name" even though it is "devoid of all forms" (*Timaeus* 50b–c). Nor does he not speak of it, as he speaks of the Good, as what is "beyond being," even though the Good is itself characterized in terms of life (*Republic* 509a).

Other strategies are thus needed to locate the traces of this *life death* in Plato's dialogues. One of them is to follow the role played by writing or inscription more generally in the Platonic corpus—its odd status as both mere supplement, that which simply adds on to speech or to thought, and necessary supplement, that which is necessary for both speech and thought, the origin, as it were, of that which it supposedly supplements. It is to show, as we saw in Chapter 5, that while Plato explicitly condemns writing for its associations with repetition, mechanical memory, artificiality, infertility, and, thus, its associations with death, he does so only by multiplying references to writing. This is precisely what Derrida demonstrates in "Plato's Pharmacy." If Plato never makes such a conception of life as *life death* explicit in any of his dialogues, what he does do, especially in late dialogues such as *Sophist, Philebus, Timaeus,* and *Statesman,* is identify a differential structure or a differential web that itself "precedes" all forms and particulars (including the form of life and everything that is called life), a differential structure in the form of an alphabet or a web, a *symplokē,* the imbrication of letters, of nouns and verbs, a community of forms or ideas such as same and other, though also, by analogy or contagion, being and becoming, the one and the many, and life and death.

In the introduction to this work, I suggested that the divided line—an image that, in the *Republic,* immediately follows the reference to the Good beyond being that I just cited (see *Republic* 509b and 509d–511e)—might be thought along these lines, that is, as a line that itself has no clear place on the line, either because it has no place at all or more than one place (in imagination, in perception . . .), a line, then, that is on the line and yet exceeds the line. But there is no need to venture into other dialogues in order to find traces of a notion of life as the drawing of lines. In Chapter 1, it became clear that *diairesis* itself, as the "art" or "science" of drawing lines between the sciences and the non-sciences, and then lines within those sciences on the basis of distinctions such as soul and body, or life and death, cannot itself be located on one side or the other of the lines it draws. There would thus be the lines and distinctions made by *diairesis,* the various lives distinguished by *diairesis,* and then there would be *diairesis* as the art or the process of drawing lines—at once active and passive, artificial and natural, human and beyond the human (because, as we saw, the lines seem to have already been drawn in the Age of Kronos).

But there is still another way to begin to approach this question of *life death* in Plato, one that is just as internal to the dialogue that I have been reading here. That is to look at those places in the Platonic corpus where life and death, or the immortal and the mortal—which are not quite the same thing—seem to be related or woven together in some way. As we saw in Chapter 6, the *Statesman* ends with the claim that the activity proper to statesmanship consists in weaving together human and divine bonds, the mortal and the immortal. Statesmanship is defined as the art of weaving together opposing virtues in order to make of them "a smooth and . . . well-woven fabric, and then entrusting to them in common for ever [*aei*] the offices of the state" (310e–311a). "Forever," says the Stranger, just as dialectic in the *Phaedrus* is said to be capable of yielding seeds in other minds and of "continuing the process for ever [*aei athanaton*]" (277a). But how are we to think this? While it is easy, at least after Plato, to think of an immortal soul living "for ever" independently of a mortal body, how are we to think of a mortal body or a human state of affairs that does not simply share in the divine *for a time*—that is, for the time it is alive—but that lives "for ever"? How are we supposed to think such mortal-immortal things in the dialogue, mortal things that share in immortality? Finally, why did Plato try to think such a thing and what does this tell us about *life death*?

In Chapters 1 and 2 we noted that, for Plato, a *zōon* was a living thing, a composite of soul and body that, due to the latter, was necessarily mortal (*thnēton*). So close is this identification of *zōa* with *thnēta* that, as we saw, the two words are sometimes used more or less synonymously. We also noted, however, that there are a couple of remarkable exceptions to this general rule. In some dialogues, the gods seem to be an exception to the rule that all *zōa* are bodily things that come into life, grow up and grow old, and eventually die. In the *Phaedrus*—yet again the *Phaedrus*—Socrates argues that if a *zōon* "is not immortal by any reasonable supposition," one might nonetheless "imagine an immortal being [*athanaton ti zōon*] which has both a soul and a body which are united for all time [*ton aei de chronon tauta xympephukota*]" (*Phaedrus* 246c). That is what a god would be, Socrates speculates, an *athanaton zōon*, a compound of body and soul that is forever united—a contradiction in terms that wreaks havoc on all the conceptual distinctions that Plato's dialogues work so hard to establish. In *Timaeus* and the *Laws*, the stars pose a similar problem as the gods.[5] How can any *zōon* live for ever?

Plato himself saw, of course, the enormous problem—though also the great potential—in affirming this notion of a god or a star as an embodied being that is nonetheless immortal.[6] It is hard to imagine these things without bodies insofar as they are visible, insofar as they appear, and yet it is just as hard to imagine them dying even though they have bodies.[7] And the same would be true, and perhaps even more true, it seems, for the universe. In the *Statesman*, recall, the universe is explicitly called a *zōon*—a *zōon* that never seems to die or perish, an immortal *zōon*, the one *zōon* (besides the gods and the stars) that continues to exist in time while all the other *zōa* within it grow old and perish. Before concluding, then, I wish to return one last time to the myth that has animated these pages from beginning to end. I do this in order to look at how Plato, in the myth of the two ages, argues for the immortality or the deathlessness of that mortal thing called the universe or cosmos. Once we do that, we will perhaps be better prepared to ask one last time about the relationship between life and life itself, or else death and life, life and everything opposed to life.

As we saw, the universe is described in the *Statesman* as a *zōon*, "a living creature [*zōon on (to pan)*] . . . endowed with intelligence by him who fashioned it in the beginning" (*Statesman* 269c–d). But because, as we recall the Stranger arguing, "absolute and perpetual immutability is a property of only the most divine things of all, and body does not belong to this class" (*Statesman* 269d), the universe as a compound of body and soul is not immutable. It is immortal but not immutable, a disjunction that borders on the nonsensical. The *Statesman*'s ingenious solution to this conundrum is to posit a perpetual alternation between an age of corruption and degeneration and an age that reverses that corruption and degeneration by reinfusing the universe with immortality or, perhaps it would be better to say, life everlasting. The universe would not be the only immortal or imperishable *zōon* (the gods and the stars, as we just saw, have a similar status) but it is surely the most spectacular. It is a *zōon* like all the other *zōa* it contains insofar as it is a mutable thing that lives in time. And yet, unlike all these other *zōa*, the universe lives in time for all time—that is, *for ever*. It is a curious thing, this *zōon* of *zōa*, this *zōon* that completely outlives all the *zōa* it contains.

Plato seemed to believe that all living things within the universe, everything but the stars, perhaps, would degenerate and perish but that the universe itself, even though it is a *zōon*, would keep on living for ever. The universe or cosmos is thus an exception to the rule of life. And yet it is an exception that also seems to provide the rule for all other living beings. For we must recall that there is a share of immortality or a trace of divinity in all *zōa*. As we saw in Chapter 1 and then again in Chapter 7, the soul is what makes a body live and move; that is the case for all *zōa*, all mortal *zōa*, though it is especially the case for man. As we have seen, there is a hierarchy of mortal *zōa* that runs from plants to animals—that is, to *zōa* in the restricted sense we saw in Chapter 1—to man or *anthrōpos*, a hierarchy that is organized, it seems, according to a scale of what is more or less divine, closer to or further away from the gods. The human, too, then, is a living being, a *zōon*, but a *zōon* of a quite different and very special kind. He is, as the Stranger of the *Statesman* puts it in the myth of the two ages, a *zōon* of a "different and more divine nature [*anthrōpoi, zōon on heteron theioteron*] than the rest" (*Statesman* 271e). In the analogy drawn by the

Stranger, what the gods were to man in the Age of Kronos, man is to the animals in the Age of Zeus. For during the age of Kronos, "God himself was their shepherd, watching over them, just as man, being an animal of different and more divine nature than the rest, now tends the lower species of animals" (*Statesman* 271e). While there is an element of immortality in all *zōa*, this immortal or divine element is most developed in man. In the *Symposium*, Diotima speaks of a striving for immortality that is to be found in all animals but that is most pronounced in man. "It is a divine affair," she says, "this engendering and bringing to birth, an immortal element in the creature that is mortal [*touto (hē kuēsis) en thnētōi onti tōi zōōi athanaton enestin*]" (*Symposium* 206c).

Every *zōon*, every living creature, is thus a compound of body and soul, which is what accounts for both its being mortal *and* its striving for immortality. The principal activity of the statesman is, it seems, to imitate and so reproduce within the state this initial blending or weaving together of mortal and immortal elements. His task is thus not unlike the one assigned by the Demiurge of the *Timaeus* to the other gods as they create other living beings, other *zōa*, in imitation of the Demiurge's creation of the universe and the gods themselves:

> turn yourselves, as Nature directs, to the work of fashioning these
> living creatures, imitating the power showed by me in my generating
> of you. Now so much of them as it is proper to designate "immortal,"
> the part we call divine which rules supreme in those who are fain to
> follow justice always and yourselves, that part I will deliver unto you
> when I have sown it and given it origin. For the rest, do ye weave
> together the mortal with the immortal, and thereby fashion and
> generate living creatures [*zōa*]. (*Timaeus* 41c–d; see 42e)

When read in the light of passages such as this, the relationship between the two ages in the *Statesman* myth can now be understood in at least three different ways—all of them legitimate and none of them completely incompatible with any other. First, it can be read in terms of just such a weaving together of mortal and immortal elements. While it would be too simple to say that the Age of Kronos is an age of immortality and the Age of Zeus an age of mortality, or that the Age of Kronos is an age of *zōē* and the Age of Zeus an age of *bios*, too simple to say that the soul is to be thought in relation to *zōē* and the soul compounded with the body in relation to *bios*, too simple to say that this troubled relationship between life and life, what

is called life and life itself, is reducible to the relationship between two gen-
erations of fathers, the Olympian Zeus and the Titan Kronos, there is, it
seems to me, the suggestion, the temptation, of just such a separation or
relation in Plato. There is the suggestion, the temptation, to separate life
from life, to supplement life with life, to say that there is real life and then
the life that is lived, life itself—life in the singular—and then the many
lives that are lived, pure life without death or a life that overcomes and reverses death and then the life that is always woven together with death, a
life before the law and then a life that must always negotiate with law, a
life, a model life, and then the many lives—the many simulacra—modeled
on that life. The Age of Kronos would thus remain, on this reading, a
model to be imitated, the singular rule or rule of a single ruler that remains
a model for the many statesmen in the Age of Zeus.

There is no denying this "two life theory," as it were, in Plato's dialogues;
as we saw in the previous chapter, there is what is called life (usually *bios*)
and then there is an elevated version of life (usually *zōē*) that looks like life
itself; there is the Age of Zeus (an age of death and of writing, an age of
law and, especially, of written law) and there is the Age of Kronos (an age
of speech or of "writing in the soul," an age of language beyond or before
language, of justice before the law). According to this reading, Kronos—
like a certain understanding of the Good beyond being—would be the
source of all things in the intelligible and the sensible realms, a source of
life that is always withdrawn, a life beyond life, real life beyond what is
called life, a realm of pure life beyond or before all phenomenal life, a
life of pure being or a life beyond being itself, not a good life, then, but the
Good as the source or principle or father of life.[8] This is, to be sure, the
most Platonist reading of the relationship between the two ages—the most
Platonist but also, perhaps, on balance, the most Platonic, the one that
would find the most evidence and corroboration in the dialogues. Such a
reading of Plato would thus locate life, life itself, on the side of being, if
not the Good beyond being, leaving what is *called* life, the homonym life—
whether *bios* or *zōē*—on the side of becoming, a mere shadow life, a trace
of *la vraie vie*, which, as in Rimbaud, would always be elsewhere, always in
withdrawal, while we are in the world.

But this relation can be read differently—that is, in a first moment, in
the *opposite* way. This second interpretation became not only possible but
necessary, one will recall from Chapter 2, the moment we began to see that
the term *automatos*, as it was being applied to the Age of Kronos, was be-

ing cast in terms—and especially terms of negation—derived from the Age of Zeus. It was in this way that the life that is lived and named *life* (*bios* or *zōē*) came to mark the life that is not lived and that would purport to go beyond the name, in this way that the categories of the Age of Zeus came to mark, determine, and define the Age of Kronos. The distinctions between species that we saw sketched out in the Age of Kronos would then be, according to this reading, not the proto-traces of natural distinctions that need to be rediscovered in the Age of Zeus but projections from the Age of Zeus upon that prior, mythical age. For it is only in the Age of Zeus that we are ever able to draw the line, and ever able then to imagine or dream of a realm or an age before the line or before this time of life—that is, before difference and death. From this perspective, human *logos*, which is itself structured like a *zōē*, is itself the measure of all things, including everything in the Age of Kronos, which is supposed to be an age without or before logos.[9]

If writing is identified in the *Statesman*, *Phaedrus*, and elsewhere with absence, death, and sterility, though also with recollection and imitation, then the Age of Zeus might be understood as the *supplement* to the Age of Kronos, the supplement that actually *invents* that prior mythical age, indeed, that invents and draws the line between itself and that mythical age. Whereas the Age of Zeus is an age of death and degeneration, an age in which life must always negotiate with death, an age in which life and death and all the categories that come with them—speech and writing, the guided and the unguided, legitimacy and illegitimacy, fecundity and sterility, and so on—must mix and mingle and be woven together, the Age of Kronos would be a mythical age upon which the pure version of the best of all these terms is projected, giving us the illusion or the phantasm of an age of speech without writing, of "writing in the soul" before all language, an age of fecundity, legitimacy, paternity, and life before all generation, the age of an Ur-Father (whether Kronos or the Demiurge) before all fathers. From this perspective, all life—including that projected upon the Age of Kronos— has its origin in the Age of Zeus.[10] The son, as we saw in Chapter 5, would be the origin of the father.

What we have here, then, are two different ways of thinking the relationship between father and son or master and disciple, a father or a master who is absent and must be remembered (Kronos/Socrates) by his son or disciple (Zeus/Plato/Young Socrates), and a son or a disciple who *invents* his father and master. It is right here, however, that this second reading, this anti-Platonist reading, as it were, begins to suggest a third reading in which the very line between the Age of Kronos and the Age of Zeus, along with everything that comes along with it, emerges against the backdrop of another structure or another thinking of life, one where, as I suggested above, life is not the opposite of death but that which must be thought together with death from the very beginning. This third reading would seem to confirm Derrida's analyses in "Plato's Pharmacy." While writing is, for Plato, a supplement to "the living and breathing word of him who knows [*ton tou eidotos logon legeis zōnta kai empsychon*], of which the written word may justly be called the image [*hou ho gegrammenos eidōlon an ti legoito dikaiōs*]" (*Phaedrus* 276a), it is clear that even speech participates in "writing" insofar as it is subject to the absence and, in principle, the death of the speaker. Hence this "living and breathing" word is itself, from the beginning, a trace of a more general differential structure—that is, a trace of *life death*. It is with this third reading, then, that we might begin to think the two ages not in terms of a priority or precedence of the one over the other (where the Age of Kronos would precede, perhaps ontologically, in the order of being, the Age of Zeus, or where the Age of Zeus would precede, perhaps epistemologically, in the order of knowing, the Age of Kronos) but as themselves the effect of another thinking of life, the effect of *life death*, of life as the drawing of lines, life as *diairesis*. Neither the father nor the son would thus come first but a *life death* that, like Heraclitus's *polemos*, first distinguishes father from son, ruler from ruled, and so on.[11] All this would be another way of saying that, in Plato, there is at once Platonism, the opposite of Platonism, and that which disrupts—or deconstructs—that Platonism.[12]

This third reading allows us to rethink not only the second reading but the first—the one we assumed, always too quickly, to be simply Platonic. For in light of this third understanding of the relationship between the two ages the question becomes one of knowing whether this life in withdrawal, this life attached to a figure of the father and teacher, to some kind of Demiurge in the Age of Kronos, marks the power and fecundity of a super sovereignty, a divine sovereignty beyond all human sovereignty—one way of reading the *epekeina tēs ousias* of the *Republic*—or whether it marks a site beyond all sovereignty, a movement of life or of *life death* that itself hides in the lines it draws between the two ages, between life and what is called life, between *bios* and *zōē*, between life and lives, but also before and between life and its others, another way of taking the *epekeina tēs ousias* (or else *khōra*), a life beyond being and beyond all sovereignty and all sovereign life, a life—the one life—deferring itself and differing always from itself. The question here regards two different ways of taking life, one that would be the very essence of ontotheology and one that would mark another time and another life before all ontotheology.

To begin to "think" this *other life*, this life as *life death*, it is necessary to focus attention on the points of transition between the two ages, the places of pivot between what seemed like an age of mortality and an age of immorality, an age of death and an age of everlasting life. Recall, for instance, that the Demiurge is said to let go of the cosmos at the end of the Age of Kronos so that it would be "left to itself at such a moment [*kata kairon*] that it moves backwards through countless ages" (270a). The implication seems to be that the Demiurge lets go of the universe at the right time, at the opportune moment, neither too soon nor too late. There is a "right time," it appears, between the age when it is always the right time and the age when the human statesman and craftsmen of various kinds must work to hit the mean and find the right time, a right time for the universe to be left on its own and a right time for it to be saved and given renewed life. What is woven in the Age of Kronos seems to get unwoven in the Age of Zeus, and what is unwoven in the Age of Zeus seems to get rewoven and restored—and at just the right time—in the Age of Kronos.[13] But this would seem to suggest that the time *between* the two ages is itself thought in terms of one of the two ages, in this case, in terms of the Age of Kronos. It suggests that the Demiurge is always there to *save* the universe from sinking into "the boundless sea of diversity" (274d–e). It suggests, perhaps, that the immortal or eternal element within the mortal is never *really* in danger of being overcome by the mortal or bodily element. For the Demiurge—not just the statesman—is a weaver, as well as a shepherd, and he is there to assure the universe's survival.

But what would it mean for this transition between the two ages to take place *without* the assured presence of a divine weaver? What would it mean for it not to be assumed or assured that the Demiurge will return at the right moment? What would Plato make of *that* contingency, a moment of contingency at the heart of the cosmos but then also, by implication, at the heart of all *zōa*, all living creatures, within it? Plato could imagine everything within the universe perishing except, it seems, the universe itself, but then also, as a result, the gods, the stars, the human soul, the immortal element within every *zōon*. But perhaps we who do not think quite the same way about the universe no longer have that luxury—or that problem. Is the Platonic notion of the Good—or of life—compatible with a universe that is not permanent, a universe that can, that will, sink into the "boundless sea of diversity"? However we answer this question, Plato's *Statesman* gives us the chance to think this moment of radical contingency within the universe and within life; it gives us, in its interstices, in the lines that are drawn and then withdrawn, this thought of a life before conception as a life that is never assured.

We saw in Chapter 2 the opening to such a thought in Plato's enigmatic use of the word *automatos* to describe not only life during the Age of Kronos but the movement of the cosmos itself during the Age of Zeus, a movement that would be not only *without* the guiding and reassuring presence of the Demiurge but free and spontaneous, autonomous, the mark of a *zōon* that lives on its own *for a time.* That *possibility* alone was enough, I tried to suggest, to threaten and destabilize everything. By untethering the cosmos from the Demiurge, by separating the Age of Zeus from the Age of Kronos, everything risks getting unhinged and going under with it: the son no longer the offspring of the father, the disciple no longer the imitation of his master, writing no longer beholden to speech, in short, paternity no longer the rule of the day.

While Plato seems tempted by this other thinking of life, this thinking is quickly consumed or recovered by another logic that reroots human life in an ideal, divine life that goes beyond human life, and that compels us to translate or to think *automatos*, the automatic movement of the universe, not as *of its own accord, freely,* or *spontaneously,* but as *automatically, without forethought, intelligence, or guidance*—that is, as merely the negation of everything that Plato values. In other words, Plato in the *Statesman* seems to turn *automatos* yet again in the direction of what is without external guidance but also without internal guidance, without a father or a teacher but also without some internal principle of life, except insofar as it imitates or recollects the principle of life itself. While Plato seemed tempted, or while we were tempted in our reading of Plato, by another thinking of life, whether his own or one inspired by, say, Hesiod or Heraclitus, a thinking of life *untethered* from its source, a life principle—though, this time, a principle of finitude as well—located in living beings that takes over at the moment beings are cut off from their source, he will have ultimately decided against it and reined in everything having to do with this other thinking of life.

Everything, that is, except the word *automatos* itself and everything it inflects or sets in motion on both sides of the pivot or the turning point we have been tracking. For this word remains in Plato's dialogue; it remains—the name of a possibility and a problem, a promise and a threat—and everything still revolves around it. Unable to translate this term without ambiguity or ambivalence, it remains a site for thinking, a place where so many oppositions rise to the surface, teeming, as it were, around it, even if, at the end of the day, Plato will have chosen, turned it in one direction rather than the other. Fundamentally ambivalent, it remains to be read as a kind of *undecidable*, a pharmakonal word, in the Platonic text, a word that marks the threshold between inside and outside, spontaneous, internal life or motion and external causality, natural, self-generated growth and technical, artful production, in short, the Age of Kronos and the Age of Zeus, though also, by the very end of the dialogue, speech, the spontaneous production of speech and science on the part of an originary lawmaker, and writing, the recollection or imitation of an original spoken word through the artifice and supplement of hypomnesic memory and writing, in other words, a certain thinking of the organism as a spontaneous life-source and a thinking of the automaton as puppet or machine without its Demiurge. Between spontaneity and automaticity, an internal and an external principle, between order and chance, the organic and a certain mechanicity, the word *automatos* divides at the origin in order to suggest two different thoughts of the origin, a dehiscence that turns the word into itself and its opposite, into itself and its homonym, into itself and the imitation of itself, the origin, perhaps, of the two ages themselves, those of Zeus and Kronos, and everything that comes so spontaneously—or automatically—along with them.

Despite everything, then, the word *automatos* remains in the dialogue, a trace, perhaps, of Plato's hesitation regarding another thinking of life, nothing less, in short, than another philosophy, a philosophy of contingency, philosophy as it *might have* become, a thinking of life untethered from its source, separated from true life or from a form of life, a life able to live, then, on its own for a time, a Platonic thinking of life within the Platonic dialogues that will have perhaps become, by whatever accident or necessity, untethered from Plato or at least from Platonism, set adrift from its father and teacher, unable to be taken in and boarded by Christianity, for example, left to wander like a phantom ship or like the universe itself in the Age of Zeus but without any promise or prospect of a god or an Age of Kronos returning to save it.

*   *   *

As the father says to his son at the beginning of Don DeLillo's *Zero K* (2016), a novel about life and immortality, time and its reversal, and wonder and philosophy, "everybody wants to own the end of the world."[14] What would it mean, then, to give up on this desire? That is the question Plato's *Statesman* continues to hold in suspense.

ACKNOWLEDGMENTS

This book emerged out of one seminar that never took place and two that did. I first began working in earnest on Plato's *Statesman* for a seminar I planned to co-teach in the spring of 1999 at the International College of Philosophy in Paris with my friend and colleague, David Farrell Krell. In the weeks and months leading up to that seminar, David and I talked incessantly about the *Statesman* and Plato more generally. Unfortunately, David could not join me in Paris that spring and the seminar had to be cancelled. More than a dozen years later, I had the chance to return to the *Statesman* and to the theme of life in Plato when I received a generous invitation from Sara Brill to give a weeklong seminar on the topic at the 2012 Collegium Phaenomenologicum in Città di Castello, Italy. A graduate seminar at DePaul in autumn 2015 gave me the chance to rethink and refine many of the ideas first presented at the Collegium. This work benefited enormously from the many questions, comments, and insights I received from both participants at the Collegium and students at DePaul. As Plato likes to say, citing Homer (*Iliad* 10.224), "when two go one sees before the other" (see *Alcibiades II* 139e), and at both the Collegium and DePaul I was oftentimes led to see what others had seen well before me.

An early version of Chapter 2 was first delivered in November 2014 at a conference at Boston College organized by John Sallis and subsequently published under the title "From Spontaneity to Automaticity: Polar (Opposite) Reversal in Plato's *Statesman*," in Plato's *Statesman: Dialectic, Myth, and Politics*, ed. John Sallis (Albany: SUNY Press, 2017), 15–31. Unfortunately, that volume appeared when this work was already in press. I was thus unable to take advantage of the many fine essays within it. Parts of Chapter 3 were published under the title "Plato's State as Exception: Foucault's Exclusion of Pastoral Power from the Dialogues," in *theory@buffalo* 15 (2011): 49–75. A first version of Chapter 5 was first presented at a conference on the *Phaedrus* organized by Burt Hopkins and Nick Pappas at City College of New York in March 2016. Finally, Chapter 7 was first presented, at the invitation of Ryan Drake and Anne-Marie Bowery, during a special session of the Ancient Philosophy Society on October 8, 2015 at the annual meeting of the Society for Phenomenology and Existential Philosophy, and then in March 2016 at the University of Maine, Orono, thanks to the generous invitation of Ciarán and Donncha Coyle, and at Memorial University in Newfoundland thanks to Peter Gratton. Last, I would like to thank my dear friend Darrell Gibbs for his expert proofreading of this work.

INTRODUCTION: PHILOSOPHY'S GIGANTOMACHIA
OVER LIFE AND BEING

1. David Farrell Krell, *Daimon Life: Heidegger and Life-Philosophy* (Bloomington: Indiana University Press, 1992), 18. This is Krell's conclusion of an analysis of Heidegger. Here are the immediately preceding lines: "[Heidegger] proceeds to identify *zōē* with *physis*, 'pure upsurgence' (103), defining the ever-upsurgent (*to aei phuon*) as 'the constantly living,' the bearer of life-fire, *to aeizōon* (104). Yet the identity of *zōē* with *physis* implies the essential proximity (*Wesensnähe*) of life and being as such, 'throughout the entire history of Western thought,' commencing with Plato and Aristotle and culminating in Leibniz and Nietzsche. In a word, *zōē* is a word for being . . ." This book has been inspired from start to finish by David Krell's work in *Daimon Life* and elsewhere on the theme of life in the Western philosophical tradition. But it has also been influenced by Krell's work on the *Statesman* itself, which has been a touchstone for Krell in so much of his work.

2. This fragment from Nietzsche's *The Will to Power*, trans. Walter Kaufmann and R. J. Hollingdale (New York: Random House, 1967), has a long and fabled history. The fragment (no 582 = KSA 12: 2[172], p. 153) is central to Heidegger's reading of Nietzsche in the third volume of his *Nietzsche, The Will to Power as Knowledge and as Metaphysics*, trans. Joan Stambaugh, David Farrell Krell, Frank A. Capuzzi (San Francisco: Harper & Row, 1987), see page 40. It is also central to Derrida's reading of both Nietzsche and Heidegger, and especially of Heidegger's reading of Nietzsche, in Derrida's unpublished seminar of 1975–76, "Life Death."

3. Jacques Derrida, *H. C. for Life, That Is to Say . . .* , trans. Laurent Milesi and Stefan Herbrechter (Stanford: Stanford University Press, 2006), 87; *H. C. pour la vie, c'est-à-dire . . .* (Paris: Éditions Galilée, 2002), 77–78.

4. The other use of the term *gigantomachia* in Plato's dialogues is found in the *Republic*, as Socrates is laying out what stories are to be told in the state and what are to be prohibited. The future guardians, he says, must be convinced not to "make battles of gods and giants [*gigantomachia*] the subject for them of stories and embroideries" (*Republic* 378c).

5. Unless noted otherwise, I have used the Loeb Classical Library translations of the Platonic dialogues, including, and especially, Harold North Fowler's translation of the *Statesman* (Cambridge: Harvard University Press, 1925). I will also occasionally cite the translations of Eva Brann, Peter Kalkavage, and Eric Salem (Newburyport, Mass.: Focus Publishing, 2012), hereafter abbreviated BKS, and Seth Benardete (Chicago: University of Chicago Press, 1986), hereafter abbreviated SB.

6. The noun *to zōon* is sometimes written *to zōion*—that is, with an "i" to mark the iota subscript (on the first vowel, the omega). Throughout this work, I will use the former spelling because it makes more apparent the word's affinity with English words such as *zoology*, *zoontology*, and so on.

7. I use the term "life itself" in a way that is quite different than—but perhaps not completely unrelated to—the way it is currently being used in contemporary debates and discourses surrounding "new materialism." This work makes absolutely no claim to contribute to those debates or discourses, though it does hope to provide some background to the question—the very old question—of life that animates them.

8. Mitchell Miller, *The Philosopher in Plato's "Statesman"*, together with "Dialectical Education and Unwritten Teachings in Plato's *Statesman*" (Las Vegas: Parmenides Publishing, 2004), hereafter abbreviated MM; Stanley Rosen, *Plato's "Statesman": The Web of Politics* (New Haven: Yale University Press, 1995), hereafter abbreviated SR; David A. White, *Myth, Metaphysics, and Dialectic in Plato's "Statesman"* (Burlington, Vt.: Ashgate Publishing Company, 2007).

9. As Rosen writes, "the highest form of political existence, ironically enough, turns out to be the transpolitical existence of *phronēsis* or, in other words, philosophy" (SR 190).

10. Frédéric Worms, "Le nouveau problème du vivant et la philosophie française contemporaine," in *La philosophie en France aujourd'hui*, ed. Yves Charles Zarka (Paris: Presses Universitaires de France, 2015), 231–247. Worms assembles an impressive list of scholars in all these domains, from the sciences (Jean-Claude Ameisen, Jean-Pierre Changeux, Philippe Descola, Thierry Hoquet, Frédéric Keck, Dominique Lestel, Catherine Malabou, Alain Prochiantz, and Francis Wolf), to politics (Marc Crépon, Michel Foucault, Guillaume Le Blanc, Claire Marin, and Pierre Zaoui), to metaphysics (including those who have criticized the philosophies of life, Alain Badiou, Emmanuel Levinas, Paul Ricoeur, and those who have tried to rethink them, Gilles Deleuze, Michel Henry, among others), to rereadings of the history of philosophy on the question of life or the living (Renaud Barbaras, Etienne Bimbenet, and Jacques Derrida, among others). One might also think of the tremendous influence of work in the life

sciences upon so many of these philosophers. See, for example, Michel Foucault's 1978 Introduction to George Canguilhem's *The Normal and the Pathological* (New York: Zone Books, 1991), 7–24, where Foucault makes a compelling case for the centrality of Canguilhem's rethinking of the concept of life in terms of error and discontinuity for so much of French philosophy in the last half century.

11. My friend and colleague at DePaul, Sean Kirkland, has argued for and demonstrated this point better than anyone in his *The Ontology of Socratic Questioning in Plato's Early Dialogues* (Albany: SUNY Press, 2012). See especially his Introduction, "Socrates and the Hermeneutics of Estrangement," xv–xxiv.

12. "Plato's Pharmacy," in *Dissemination*, trans. Barbara Johnson (Chicago: University of Chicago Press, 1981), 61–171; *La dissémination* (Paris: Éditions du Seuil, 1972), 69–197; first published in *Tel Quel*, nos. 32 and 33 (1968); hereafter abbreviated "PP." with pages references to the English edition and then the French.

13. In order to distinguish, for example, the faculty of imagination from belief, and then both of these faculties of the visible realm from the two faculties of the intelligible realm, mathematical thinking and dialectical thought, Plato places all these faculties and their corresponding objects along a line, a visible line that is itself subdivided according to certain proportions in order to distinguish and hierarchize these faculties and their objects (see *Republic* 509d–510a).

14. Levinas famously cites this line from Rimbaud at the beginning of *Totality and Infinity*. See *Totality and Infinity*, trans. Alphonso Lingis (Pittsburgh: Duquesne University Press, 1969), 33.

15. Derrida sometimes writes this with a dash (*la vie–la mort*) and sometimes without (*la vie la mort*).

1. THE LIFELINES OF THE *STATESMAN*

1. From Derrida's introduction to *Parages* (Paris: Éditions Galilée, 2003), 12; translated by John P. Leavey in *Parages* (Stanford: Stanford University Press, 2011), 4: "where does the line of parti(cipa)tion pass between the event of an inaugural statement, a citation, a paraphrase, a commentary, a translation, a reading, an interpretation?"

2. Mitchell Miller argues that much of the drama of the *Statesman* consists in the Stranger's education of the Young Socrates, his attempt to see whether Socrates has "a spiritual heir in the younger Socrates, someone who will incarnate the Socratic essence after he is gone" (Miller, *The Philosopher in Plato's "Statesman"*, together with "Dialectical Education and Unwritten Teachings in Plato's *Statesman*" (Las Vegas: Parmenides Publishing, 2004),

7; hereafter abbreviated MM). The Stranger "must somehow awaken the Socratic essence within Young Socrates, enabling him to live up to his name and become genuinely Socratic" (MM 15). Miller thus follows the development of the Young Socrates from being "energetic" but "overconfident" and "uncritical" (MM 7) to being cognizant of his own ignorance (MM 58). By the end of the dialogue, he will have "recognized his own lack of *epistēmē*" and so "attained the genuinely Socratic station between ignorance and wisdom, the knowledge of ignorance" (MM 103). The goal of the dialogue is thus the education of Young Socrates, or, on another level, the reader or listener, or else the young Academician, who, on Miller's account, is supposed to come to self-recognition by bearing witness to the Young Socrates's many failures and ultimate insights (MM 116).

3. David White argues that "Plato wanted to speak through a Stranger because things which had been relatively clear to the Socrates of the middle period were clear no longer—that is, the philosophically familiar had become 'strange,' at least to a certain extent" (David A. White, *Myth, Metaphysics, and Dialectic in Plato's "Statesman"* [Burlington, Vt.: Ashgate Publishing Company, 2007], 8; hereafter abbreviated DW). Miller argues that because "the true Parmenidean spirit is not well represented by the eristics" (such as Zeno), Plato "proposes the Stranger instead as a true heir to Parmenides" (MM 13).

4. As Stanley Rosen puts it, "Plato is neither the Stranger nor Socrates, but both and more. The two men represent finally irreconcilable aspects of the philosophical nature" (Rosen, *Plato's "Statesman": The Web of Politics* [New Haven: Conn.: Yale University Press, 1995], 77; hereafter abbreviated SR). I agree with the first claim, but not necessarily the second.

5. Because of the presence of Theodorus of Cyrene, the god Ammon is briefly evoked at 257b. Identified with Zeus, Ammon was the great god of Egyptian Thebes, the chief god of the Egyptian pantheon. *The Oxford Classical Dictionary* tells us that Greek interest in Ammon was "probably mediated through the city of Cyrene (on whose coins his head is shown from the early 5th century with the typical ram's horns)."

6. One can, of course, always speculate about why no such dialogue exists. One can, for example, argue that many aspects of the true philosopher actually emerge in the search for the sophist in the *Sophist* and the statesman in the *Statesman*, so that a separate dialogue on the philosopher was unnecessary or redundant. Or one can claim, as Mary Louise Gill has done in her work *Philosophos: Plato's Missing Dialogue* (Oxford: Oxford University Press, 2012), that Plato deliberately "withholds the dialogue in order to stimulate his audience to combine the pieces into the full portrait he did not paint" (1).

7. To appeal to a distinction that Socrates himself uses in the *Sophist*, Theodorus would essentially be comparing, putting into proportion, those who are "of no worth" and those who are "worth everything" (*Sophist* 216c; see MM 9). Miller rightly argues that while "the explicit tone is one of mutual deference, especially towards Socrates," the dialogue takes place against the backdrop of two conflicts: Socrates's conflict with the unphilosophical many of Athens, as that is made manifest in his impending trial, and a conflict between mathematics (in the person of Theodorus) or the method of mathematics, which does not question its fundamental assumptions and requires the use of figures, and philosophy proper (MM 4–5; see 34, 68–69). As Miller reminds us, the jurors at Socrates's trial are "confronted with just those issues which Socrates has posed for the Stranger in the trilogy: Who is the sophist? Who is the statesman? Who is the philosopher?" (MM 2)

8. Another way of asking this same question, this time in reference to the divided line of the *Republic*, which I evoked in the introduction and will return to in the conclusion, is to ask whether imagination and belief really can appear on the same line as mathematical thinking and dialectic—that is, whether the former really are *proportionate* to the latter in any way.

9. Some commentators, including Miller, attribute the very last lines of the dialogue to the elder Socrates rather than the younger one (311c). I see no compelling reason for such a sudden change in interlocutor.

10. I agree entirely with Miller's claim that the *Statesman*, despite being a later dialogue, and one in which Socrates is mostly silent, "substantively repays being read as a genuine dramatic dialogue" (MM xii). I am convinced, for example, by Miller's close account of the *pedagogical* relationship between the Stranger and the Young Socrates and his claim that the *Statesman* is an act of indirect communication between Plato and the younger generation of Academicians" (MM 6). I am less convinced than Miller about "the significance of the elder Socrates's self-imposed silence throughout the Stranger's and Young Socrates's inquiry" (MM xv; see xxv, xxxiii, 8); that is, I am not so certain that Socrates is "conspicuously present in his silence" (MM 8).

11. Rosen agrees with this assessment: "It is certainly not true that *diairesis* is useless in these studies. All thinking, even the contemplation of pure form, includes *diairesis* or the separation of kinds with respect to like and unlike" (SR 100).

12. Rosen rightly argues that central to the dialogue is the "problem of philosophical methodology" (vii) and that, in short, "the method is the doctrine" (SR 12). Rosen later claims that the "question of philosophical method is much more visible and . . . much deeper in the *Statesman* than it is

in almost any other dialogue. Perhaps only the *Sophist* and the *Parmenides* are from this standpoint at the same level as the *Statesman*" (SR 100). This question of method is also nicely underscored throughout Melissa Lane's *Method and Politics in Plato's "Statesman"* (New York: Cambridge University Press, 1998). Here, for example, are the final words of the book, which recall and justify Lane's title: "The capacities for exemplifying and dividing, and for finding the mean, which are exercised in the Stranger's methods of inquiry, are the same capacities on which the political knowledge which it defines will have to rely. Method and politics in the *Statesman* become one" (202).

13. As Charles Kahn argues, "a person is properly regarded as possessing *politikē* even if he does not exercise political rule (259a, recalled at 292e). It is natural to suppose that, in thus insisting on the claim of expertise for someone who is not actually a ruler, Plato has his own situation in mind"; see Charles H. Kahn, "The Myth of the *Statesman*," in *Plato's Myths*, ed. Catalin Partenie (New York: Cambridge University Press, 2009), 153.

14. We find this analogy not only in the *Republic* (389b–c), which would seem to precede the *Statesman*, but in the *Laws* (720a–e), which is usually thought to follow it.

15. In the *Phaedrus*, Socrates speaks of the orator's ability to make small things seem large "by the power of their words [*rhōmēn logou*]" (267b).

16. This distinction between head and hand, *kephalos* and *kheir*, is crucial to the *Laws*, where the Night Synod is compared precisely to a head (see 961e–962b) and an overreliance on the hands leads to a kind of slavishness in otherwise free men.

17. I say *quasi* here because it is unclear how to characterize or locate this rather unique ability to distinguish between various abilities, this unique ability to *draw the line*. Is the method or ability by which one draws the line between the arts and the non-arts, and then between or amongst the arts themselves, itself an art or a science? Is it instead a philosophical craft or knack of some kind? Because *diairesis* itself appears nowhere in the *diairesis* of the sciences we have just seen, because it is not grouped with the sciences or the non-sciences, it is impossible, and perhaps for good reason, to say one way or another. I will return to this point in my conclusion.

18. It is surely no coincidence, then, that Socrates compares Lysias's written discourse to an epitaph—that is, to the "inscription" on the tomb of Midas the Phrygian. Both are not only haphazardly organized, such that, as Socrates says, any line could have come before or after any other, but inorganic and moribund, in a word, lifeless.

19. See Guy Berthiaume, *Les rôles du mágeiros: Étude sur la boucherie, la cuisine et le sacrifice dans la Grèce ancienne* (Leiden: E. J. Brill, 1982). Berthiaume

argues that the *mageiros* has the triple role of sacrificer, butcher, and cook. The term appears to have become prevalent only during the second half of the fifth century, as cuisine became more refined in Athenian life. The *mageiros* was a professional who exercised his art during public rituals and festivals, at the prytaneum, and in private festivals. The function of the *mageiros* seems to have been restricted to men, even at festivals or celebrations reserved for women. Berthiaume speaks of the art of carving up the sacrificed animal "by following the articulations," a formula he takes from the *Statesman* (50–51). The *mageiros* is defined elsewhere in the dialogues as a craftsman, a *dēmiourgos*, whose work is to "slaughter and skin and, after cutting up the joints, to stew and roast" (*Euthydemus* 301c–d), to "cut up the animals and distribute the meat to the foreigners and craftsmen and their servants" (*Laws* 849d; at *Republic* 332c the task of seasoning meat is also attributed to the *mageiros*; see also *Republic* 373c; *Theaetetus* 178d; *Gorgias* 491a, 500b; *Minos* 316e, 317b; *Theages* 125c; and, finally, *Statesman* 289). Berthiaume cites the following inscription: "Kallistratos, the mageiros, dedicated to Asclepius." It is hard not to wonder whether Socrates's famous last words should not be thought in this context. As a lover of the methods of separation and collection, Socrates might be considered a master *mageiros*— that is, perhaps, in Plato's eyes, the only *true* or *real mageiros*. For an excellent analysis of separation and division in relation to the *mageiros*, see Holly Moore's "Animal Sacrifice in Plato's Later Methodology," in *Plato's Animals*, ed. Jeremy Bell and Michael Naas (Bloomington: Indiana University Press, 2015), 179–192.

20. As others have noted, *to melos* is also song or a strain, so that *kata melea* might suggest not just "limb by limb" but "strain by strain." Miller underscores the *organic* nature of Plato's image: "The Stranger compares the arts, taken as whole, to an organic totality. The various arts are compared to the various 'limbs' or 'members' of this body. As such, they are essentially interrelated, not merely in the abstract sense of being physically connected but rather in the sense of cooperating, each with each other, and contributing, each in its specific way, to the well-being of the whole" (MM 76).

21. The topic of sacrifice is a complicated one in Plato. Beyond the famous "cock" that Socrates suggests sacrificing to Asclepius at the conclusion of the *Phaedo*, there are several references to bulls being sacrificed (*Critias* 119, *Alcibiades II* 149c) and the suggestion that, in the *Phaedo*, Socrates himself, with his bull-like eyes, is the bull or minotaur that is about to be sacrificed (*Phaedo* 117b). For the question of how and what is to be sacrificed (plants as opposed to birds or animals, animate as opposed to inanimate beings, animals as opposed to human beings, bloody as opposed to bloodless sacrifices) see, for example, *Laws* 782c and 956a. Because the

process of *diairesis* is associated with the practice of dividing a sacrificial animal or *zōon* at its joints—that is, at its *diaphuai* or its *arthra*—it is perhaps worth thinking this division in relation to a passage from the *Laws* that describes a process of voting for magistrates that concludes with a final round in which votes are cast after walking through or between the slain—sacrificial—animal victims (*Laws* 753d).

22. The word used here for "withdrawal," *apēllachthai*, is the same one used at *Phaedo* 66d to describe the way in which the philosopher must withdraw from the body in order to perceive realities.

23. Plato has the Stranger make yet another methodological point at this juncture: when the Young Socrates says he agrees with the distinction that has just been made by the Stranger, the Stranger responds that when one is arguing all that is required is the agreement of the two interlocutors. "So long as we ourselves are in agreement," he says, "we need not bother about the opinions of others" (260b). We thus see that the method of *diairesis* is being carried out within or in concert with a form of dialectic that aims not necessarily at the truth of things but at the agreement of the interlocutors—yet another Platonic, indeed Socratic, trait of this dialogue.

24. We see something similar in the *Sophist* when the hunting of "lifeless things [*apsychōn*]" is opposed to the hunting of "living things [*empsychōn zōōn*]" (*Sophist* 220a) and the purification of "inanimate bodies [*tōn apsychōn sōmatōn*]" is contrasted with the purification of living, animate ones—that is, with *zōa* (*Sophist* 226e–227a).

25. This translation of *empsycha* and *zōa* tells us a great deal about translation itself and the assumptions behind it: if to translate means simply to go beyond the signifier to the signified, to strip away the body of a word, as it were, in order to arrive at its meaning or its "soul," then Fowler is surely justified in translating both words as "living objects." If translation is not, however, such a stripping away of the body of the signifier in order to reach a word's inner meaning, then Fowler's translation is perhaps already a Platonic or Platonizing translation that reveals as much about the translator's Platonism as Plato's. Tell me how you translate, it might be said, and I will tell you who you are . . .

26. In the *Phaedo*, Socrates brings humans (*anthrōpoi*) together with animals (*zōa*) and plants (*phuta*) as things that all have birth or a *genesis* (*Phaedo* 70d; see *Symposium* 188a–b). This list—or this hierarchy (humans, animals, plants)—can be extended beyond living things to non-living ones, whether natural or man-made. In the *Sophist*, plants and animals, as ensouled, animate beings, are opposed to inanimate ones: "There are all the animals, and all the plants that grow out of the earth from seeds and roots, and all the lifeless [*apsycha*] substances, fusible and infusible, that are formed

within the earth" (*Sophist* 265b; see *Timaeus* 77b). At *Phaedo* 110e, Socrates speaks about "ugliness and disease in earth and stones and animals and plants." Finally, the second section of the divided line contains "the animals about us and all plants" as well as "the whole class of objects [*skeuaston*] made by man" (*Republic VI* 510a; see also 515a and 523b–c, as well as 596c, e, and *Laws* 765e and 889c).

Interestingly, animals, *ta zōa*, are a favorite example of something to be imitated, something of which craftsman, for instance, produce likenesses (*Republic* 401b). In the *Symposium*, they also serve, and there is surely no coincidence in this, as an example of things that might be beautiful without themselves being the form of beauty (*Symposium* 211a). Finally, as living, ensouled, self-moving things, they appear along a continuum in the individual's ascent in the visible realm from animals to stars to the sun itself: "This, then, at last, Glaucon, is the very law which dialectics recites, the train which it executes, of which, though it belongs to the intelligible, we may see an imitation in the progress of the faculty of vision, as we described its endeavor to look at living things [*ta zōa*] themselves and the stars themselves and finally at the very sun" (*Republic* 532a).

27. In the *Sophist*, for example, the Stranger seems to detach animals from plants when he suggests that the sophist is someone who can "make all things," "I mean you and me among the 'all,' and the other animals [*zōa*] besides, and the trees [*dendra*]" (*Sophist* 234a).

28. As we will see, the *zōa* that will pose the most problems for Plato will be those located on the extremes, the gods, the stars, the universe, at one end, and plants on the other. On the one hand, plants seem to be alive, to participate in *zōē* like all other *zōa*. They too would thus have a soul or a form of soul, as a passage from the *Timaeus* seems to suggest; it is there said that the plant "lives [*zēi*]" and "is not other than a living creature [*oukh heteron zōou*]," even though "it remains stationary and rooted down owing to its being deprived of the power of self-movement" (*Timaeus* 77b–c). The plant is a *zōon*, then, even though it lacks one of the attributes typically given to a body by a soul, namely, self-movement. But that is really just the beginning of the problems posed by plants in Plato. For a good summary and analysis of the place of plants in Plato, see Jeffrey T. Nealon's *Plant Theory: Biopower and Vegetable Life* (Stanford, Calif.: Stanford University Press, 2016), 29–32. In a later chapter of this magnificent book, Nealon looks at Derrida's understanding of plants from *Glas* to *The Beast and the Sovereign* (49–81).

29. See *Gorgias* 483d, *Protagoras* 321b–c, 334b, *Republic* 466d, *Menexenus* 238a, *Philebus* 11b, 31d, 16a, *Laws* 644a, *Phaedo* 70d, *Phaedrus* 249e–250a, *Letter XIII* 360d, and, perhaps, *Symposium* 186a. As for places in the dialogues

where *anthrōpoi* seem to be distinguished from other *zōa*, or at least are not immediately conflated with them, see *Phaedo* 111a–b.

30. On this "art of rearing living beings [*zōotrophikē*]," see *Statesman* 261d, 263e, 267a.

31. Protagoras recount how Epimetheus "heedlessly squandered his stock of properties on the brutes [*ta aloga*]," so that when Prometheus arrived he found "the other creatures [*alla zōa*] fully and suitably provided" while "man was naked, unshod, unbedded, unarmed" (*Protagoras* 321c).

32. On the Greek/barbarian distinction, Miller recalls that Isocrates had "hoped that a common hostility to the barbarians might unite the Greeks" and that "*Plato himself* had Socrates express this general view at *Republic* 470a." According to Miller, Plato in the *Statesman* essentially "challenges this Hellenism" (MM 22).

33. Miller writes: "As with 'barbarians,' the presence of the *word* 'beasts' gives the semblance of a genuine kind; but any such semblance dissolves in the analogous case of 'all numbers other than 10,000.' In truth all three are merely negative groupings." In other words, "the sole meaning of 'beasts' is 'not-men'" (MM 21).

34. Such a division into male and female already has some precedent in the *Republic*. Socrates there bases his claim that women should be able to share in all the pursuits of men by suggesting that "natural capacities are distributed alike among both creatures [*en amphoin toin zōoin*]" (*Republic* 455d). While men and women are part of the same *genos*, the *genos* of *anthrōpos*, that *genos* seems to include two different kinds of *zōa*, male and female. From this perspective, Greek men might have more in common with barbarian men than with Greek women.

35. This method is supposed to help them pick out, track, and capture the sophist. But, says the Stranger, because the "tribe" of sophists is so "hard to catch" and "define," they will practice the "method of hunting" on something easier to catch, namely, on angling as a kind of hunting. By practicing their method in this way, they will then have a "pattern" for how to proceed (218d–221c). But angling will do more than just provide an example of the method to be used in seeking the sophist; it will put them on the path to a first definition because the sophist, like the angler, will be a sort of hunter. Throughout this search, the sophist is himself described as a "creature [*thērion*]" that is hard to grasp and whose *track* [*ichnos*] must be followed (*Sophist* 226a–b). It is thus not only the object of the search, the sophist, who is described in terms of the hunt but the very activity of the search.

36. The Stranger notes earlier in the dialogue that this method of argument is totally indifferent to the content or referent of a distinction (*Sophist* 227a). What matters is the conceptual opposition or distinction,

regardless of the value commonly attached to the things being differenti-
ated. When discussing hunting, therefore, the art of louse-hunting is just as
significant as the art of the general, even though a general is nobler than a
louse. The point is to see what is *related* and what is not in the arts. Hence
the statesman can initially be compared to the swineherd.

37. Miller's analyses of these jokes and puns (at 266a–d) are masterful
(see MM 29–33). As Miller shows, the problem with comparing the statesmen to the swineherd, and man to a pig or a featherless biped, is that man is
"revealed solely in his brutish aspect, so that the governing of man will come
to light as a sort of animal-keeping" (MM 31–32).

38. White calls the myth of the *Phaedrus* a "counterpart" to that of
Statesman (DW 10) because the charioteer analogy of this latter "is reminiscent of the mythic account of soul in the *Phaedrus*" (DW 33).

39. The Young Socrates says the Stranger has finished up the argument
well, paying the debt and throwing in the "digression [*ektropēn*]" concerning
the nature of *diairesis* for "interest [*tokon*]" (267a). Recall the importance of
this term *tokos*, at once interest and offspring, at *Republic* 507a.

2. LIFE AND SPONTANEITY

1. Charles Kahn writes: "All of the Stranger's methodological effort
(from 261 to 267) has been expended on zoological classification—that is, in
separating human beings from other animal herd animals" ("The Myth of
the *Statesman*," in *Plato's Myths*, ed. Catalin Partenie [New York: Cambridge
University Press, 2009], 154); hereafter abbreviated CK. "In contemporary
language," writes Stanley Rosen, "the Stranger fails to separate biology,
technology, and politics from one another" (*Plato's "Statesman": The Web of
Politics* [New Haven: Yale University Press, 1995], 74; hereafter abbreviated
SR). Rosen sees the problem with the *Statesman* as a kind of bio-politics,
managing the life, the biological life, of the herd. As for David White, he
argues that what we get at the end of the shorter way is the notion of the
"king as charioteer [who] rules over featherless bipeds with featherless souls,
souls which have not yet seen any reality, anything of the Forms" (*Myth,
Metaphysics, and Dialectic in Plato's "Statesman"* [Burlington, Vt.: Ashgate
Publishing Company, 2007], 33; hereafter abbreviated DW).

2. As Miller says, the move to myth is "an unlikely turn. On the face of
it, no mode of discourse could be more the opposite to rational, method-
ologically principled *diairesis*." And yet, as Miller will go on to argue, Plato
gives us a *mythos* that is "transparently *logos*" (*The Philosopher in Plato's
"Statesman"*, together with "Dialectical Education and Unwritten Teachings
in Plato's *Statesman*" [Las Vegas: Parmenides Publishing, 2004], 36; hereafter
abbreviated MM).

3. Rosen calls this myth a "bizarre fairy tale" (SR 2); he labels the Age of Kronos the "counternormal" epoch and the Age of Zeus the "normal" one (SR 41).

4. Pierre Vidal-Naquet, "Le mythe platonicien du *Politique*, les ambiguïtés d l'âge d'or et de l'histoire," in *Le Chasseur Noir: Formes de pensées et formes de société dans le monde grec* (Paris: La Découverte/Maspero, 1983), 369. Vidal-Naquet brings out many of the ambiguities in Plato's account—the role of Atreus and Thyestes in the preamble to the myth, the role of Kronos in the tale, and so on. The golden age is one of simplicity, an age followed by a pastoral life, then an agricultural one (361–362). Miller reminds us that, in Aristotle's *Constitution of Athens*, "the tyranny of Pisistratus (561–527) was often recalled and praised as *ho epi Kronou bios*, 'life under Kronos'" (MM 43).

5. As White notes, "The notion of measure is thus introduced at the outset of the myth and in a context of fundamental significance, since the relation between deity and cosmos is contoured by a given measure of cycles" (DW 38). As he later writes, "the demiurge has constructed the cosmos with an eye toward due measure: at one extreme, the rotation of the cosmos in one direction; at the other extreme, the counter-rotation of the cosmos" (DW 47).

6. *Statesman*, trans. Eva Brann, Peter Kalkavage, and Eric Salem (Newburyport, Mass.: Focus Publishing, 2012); hereafter abbreviated BKS. *Statesman*, trans. Seth Benardete (Chicago: University of Chicago Press, 1986); hereafter abbreviated SB.

7. White says we should "call this deity the demiurge, since even if this deity differs significantly from the Demiurge in the *Timaeus*, it achieves a similar formative function with respect to the origin of the cosmos" (DW 39).

8. That this is the least corruptible of bodies is explained by the fact that it "moves, so far as it is able to do so, with a single motion in the same place and the same manner [*en tōi autōi kata tauta mian phoran kineitai*]," that is, by the fact that its circular motion "involves the least deviation *from its own motion*" (269e). But this is not sufficient to explain why it is not still mortal, why it is not simply long-living or long-lasting rather than immortal or everlasting. It is, as we will see, the hypothesis of a Demiurge intervening periodically in the degeneration of the universe to infuse it with a "renewed immortality" that is put forward to resolve or explain this paradox.

This is the kind of speculation we find in Aristotle's *On the Heavens* or, for our purposes, the *Phaedrus* (see, for example, 245d–e) or, better, the *Timaeus*. For example, when it becomes a question of knowing whether they are right "in describing the Heaven as one" or whether it would be "more correct to speak of heavens as many or infinite in number," the answer comes in the form of speculation: "in order that this Creature might resemble the all-perfect Living Creature in respect of its uniqueness, for this reason its Maker made neither two Universes nor an infinite number, but there is and will continue to be this one generated Heaven, unique of its kind" (*Timaeus* 31a–b). See also *Laws* 893d–894d, where the Athenian describes various forms of motion, including rotation around a fixed point.

9. *Physics*, trans. Philip H. Wicksteed and Francis M. Cornford (Cambridge: Harvard University Press, 1970). Aristotle writes: "The etymology of *automaton* indicates this; for the expression *matēn*—'for nothing,' 'to no purpose'—is used in cases where the end or purpose is not realized, but only the means to it" (197b23–28). Or again: "So then *automaton*, as the form of the word implies, means an occurrence that is *in itself* (*auto*) to *no purpose* (*matēn*). A stone falls and hits someone, but it does not fall for the purpose of hitting him; the fall accordingly was 'in-itself-to-no-purpose'—a chance result—because the fall might have been caused by someone who had the purpose of hitting the man" (197b29–33).

10. Aristotle characterizes *tuchē* (fortune or chance) as a "special class" of *automaton*—that is, a special class of the "accidental." While the latter term can be used to describe what happens to animals or inanimate beings, the former is used only for beings with self-direction, beings capable of purpose or choice (*proairesis*)—that is, human beings (197a36–197b7). One can thus speak of things happening accidentally to animals or inanimate beings, but one can speak of good fortune or ill fortune happening only to human beings. "It is clear, then, that when *any* causal agency incidentally produces a significant result outside its aim, we attribute it to *automaton*; and in the special cases where such a result springs from deliberate action (though not aimed at it) on the part of a being capable of choice, we may say that it comes by *tuchē*" (197b19–23). Aristotle's understanding of *automaton* is thus not incompatible with Plato's: It has to do with what happens accidentally, without or outside the aim of a causal agency. (See also *Metaphysics VII* for the relationship between *automaton* and production.) Lacan picks up on this distinction between the two terms in "Tuchē and automaton," in *The Seminar of Jacques Lacan*, ed. Jacques-Alain Miller, Book XI, *The Four Fundamental Concepts of Psychoanalysis*, trans. Alan Sheridan (New York: Norton, 1977). Lacan there identifies *tuchē* with the encounter with the real, an essentially missed encounter with the site of trauma, an origin that appears accidental (see 52–54, 63).

For a fascinating reading of *automatos* in Aristotle, and not just in *Physics* but also in *Generation of Animals*, see Emanuela Bianchi's *The Feminine Symptom: Aleatory Matter in the Aristotelian Cosmos* (New York: Fordham University Press, 2014), particularly the chapter "Necessity and *Automaton*,"

51–84. As Bianchi argues, *automatos* vacillates in Aristotle between the automatic movement of highly technical, artfully constructed machines and the spontaneous, a-teleological motion or growth of organic beings. Bianchi writes, for example: "Aristotle's notion of *automaton* carries with it a schematized rendering of these contradictions—on the one hand that which causes wonder, is marvelous and illusory but nonetheless ultimately designed and therefore reasonable, teleological, and knowable; on the other that which proliferates senselessly, and appears as disruptive, aleatory, and going nowhere. *Physics* II.6 gives an instructive illustration of this ambiguity" (73). And this ambiguity *in* Aristotle gets reproduced, as Bianchi shows, in the history of commentaries *on* Aristotle, which have gone back and forth from the beginning to this day on the question of whether, for example, *automatos* is to be understood as an internal or an external cause.

11. Rosen makes the same point; while the Age of Kronos can initially look like "a film that we are rewinding on our video recorders," "this image is not adequate to the details in the Stranger's account" (SR 43; see 54).

12. Stanley Rosen points out that Kronos (or the Demiurge) here resembles Zeus at *Phaedrus* 246e, driving a winged chariot around the heavens and "arranging all things and caring for all things" (SR 54). This is yet another of the many connections between the *Phaedrus* and the *Statesman*.

13. Hesiod, *The Homeric Hymns and Homerica*, trans. Hugh G. Evelyn-White (Cambridge: Harvard University Press, 1943).

14. For an excellent analysis of this aspect of the dialogue—that is, for the possibility of talking, philosophizing animals—see David Farrell Krell's essay "Talk to the Animals," in *Plato's Animals*, 27–39.

15. As for this natural or innate desire, White contrasts the position of Scodel, for whom the innate desire of the cosmos is "its desire to actualize the potential for life it has been given by its maker," and his own, which he derives in part from a reading of *Phaedo* 75b. In that passage, "Socrates says of perceiving equal objects that 'before we began to see or hear or otherwise perceive, we must have possessed knowledge of the Equal itself if we were about to refer our sense perceptions of equal objects to it, and realized that all of them were eager [*prothumeitai*] to be like it, but were inferior.' The desire in question is not for life, as Scodel maintains, but for remaining alive by exclusively retaining the specific formal nature which had been firmly in place during the Kronos cycle, just as, according to the *Phaedo*, instances of equal things are 'eager' to be as like the Form equality as is possible for such embodied entities" (DW 228–229n11). A quasi-synonym of *moira*, *heimarmenē* is used mostly in later dialogues (see *Phaedo* 115a; *Laws* 873c, 904c; *Gorgias* 512e; *Theaetetus* 169c; *Republic* 619c; *Timaeus* 89c).

16. White rightly recalls that this "appeal to remembering as the source of this identity recalls the doctrine of anamnesis so central to the *Meno* and *Phaedo*, which in turn evokes the object of anamnesis—the Forms" (DW 49).

17. This phrase is repeated at *Menexenus*, where it is said that women "imitate" the earth or the land (*gē*)—not the other way around—"in the matter of conception and birth [*kuēsei kai gennēsei*]"; in other words, the women of Athens bear and nourish their offspring by imitating the way the earth bears and nourishes her offspring (*Menexenus* 237a–238a).

18. Miller speaks of self-relation in a similar way: "Precisely as different from the constant and self-same god, the cosmos—and man within it—must differ from itself" (MM 109). This means that higher elements in the human and in the human soul must rule over lower or inferior ones.

19. See CK 149.

20. The question, in short, is how to account for what appears to be an ambiguity, if not a fundamental ambivalence, in Plato's use of the term *automatos*. One way to avoid the problem is simply to posit, like Stanley Rosen, "two kinds of spontaneity," one in the Age of Kronos, a spontaneity that refers to "motions independent of human intention or work," and one in the Age of Zeus, a spontaneity that refers to "natural motions, cosmic and human," and to claim that only the latter "could have political significance" (SR 61). While I find it implausible that Plato would use, in the space of just a few lines, the very same term in such different or even opposing ways, and even more implausible that such very different uses would not, quite apart from Plato's intentions, contaminate and thus complicate one another, Rosen has at least seen the problem with the term *automatos* and proposed a solution to it. That is more that can be said for the vast majority of commentators on the dialogue.

21. Later in the *Cratylus*, the term *automatos* is used to characterize the view that names do not have a natural correctness but are distributed "haphazardly," that is, at random (397a).

22. *Diogenes Laertes*, trans. R. D. Hicks (Cambridge: Harvard University Press, 1970), IX.5.

23. In the *Iliad*, too, this enigmatic word seems to waver between negative and positive valences, between, for example, "unbidden" or "uncalled" (2.408–409) and "of their own accord," "self-bidden," "not unwillingly" (5.748–750), "on their own," "of themselves," and so on (18.372–377). See my "Le tournant de l'amitié chez Homère," in *Tympanum 4: Khoraographies for Jacques Derrida*, July 15, 2000; online journal at http://www.edu/tympanum/4/.

24. See also *Republic* 520b where the philosophers of most cities are said to grow up "spontaneously," that is, unguided, uninstructed. The word

automatos does appear to be positively valued in a passage in the *Republic* where Plato wants to oppose the spontaneity or non-artificiality of dialectic to the contrivances and artificiality of rhetoric. Glaucon says that the multitude has experienced "only forced and artificial chiming of a word and phrase, not spontaneous and accidental [*automatou*] as has happened here" (*Republic* 498d–e). The suggestion seems to be that an argument is more convincing when it develops naturally, of its own accord, rather than being artificially controlled—perhaps even written—from the outset. Dialectic would thus be spontaneous inasmuch as it is not planned or written, even if it is, as in the *Republic* itself, guided or oriented by one of the interlocutors.

25. A reading of the *Statesman* in light of the *Sophist* might therefore conclude that those things that are said to come up spontaneously in the Age of Kronos—that is, naturally, *automatos*, without human art—are, in fact, the result of this "divine art." As we are about to see, however, the term typically suggests not that which happens *with* divine art but that which happens *without* any art at all.

26. Stanley Rosen notes that, in the *Statesman*, "Eros is entirely subordinated to politics" (SR 189), so subordinated, in fact, that it is not even mentioned in the dialogue (SR 154). Rosen concludes that "the Stranger differs importantly from Socrates in his lack of interest in this topic" (SR 154). Whereas "the Stranger exhibits no interest whatsoever in Eros," "Socrates claims to understand nothing else" (SR 2).

27. Jacques Derrida, "Plato's Pharmacy," in *Dissemination*, trans. Barbara Johnson (Chicago: University of Chicago Press, 1987), 80–81

28. The notion of a world-soul in *Timaeus* will seem to settle many of these problems, but then the language of *automatos* will have also disappeared.

29. Charles Kahn points out several similarities between *Statesman* and *Timaeus*: "the notion of the cosmos as an intelligent living creature constructed by a craftsman god (*dēmiourgos*), with innate circular movement as the closest a changing body can come to the unchanging stability of divine beings (269d); also, a great disorder preceding the creation of the cosmos (273b)" (CK 151). And then there is the "destruction of mankind by fire and water (*Timaeus* 22c, echoed in *Critias* 112a and *Laws* 677a)" (CK 151).

30. Insofar as Cratylus was himself a Heraclitean, it is hardly surprising that Heraclitus would be so prominent in the dialogue. Even if much of the dialogue consists of a sort of comically enacted performative contradiction between the doctrine of natural correctness that Socrates seems to be proving and the demonstration itself, nothing in the dialogue contradicts the proposition uttered early on that when we use names "we teach one another something, and separate things according to their natures" (*Cratylus*

388a). As the dialogue goes on to show, the dialectician must nonetheless supervise the name-giver, just as, in an analogy that will become important when we return to the *Statesman*, the steersman or *kybernētēs* must supervise the carpenter who makes the rudder, the steersman thereby steering the one who makes the instrument with which to steer the ship (*Cratylus* 390d). Following Socrates's analogy, the dialectician is a sort of helmsman capable of steering the one who makes the instruments with which we steer ourselves.

31. Aram Frenkian, *Études de Philosophie présocratique, Héraclite d'Ephèse*, (Cernauti: Glasul Bucovinei, 1933). Other commentators, and most notably Charles Kahn, have argued that there is also an Empedoclean influence on the myth (see CK 150–151). Among these Empedoclean elements are the "symmetrical movement back and forth between two diametrically opposed situations" (marked in Greek by the *tote men . . . tote de* structure found at *Statesman* 269c and 270a; Kahn also sends us to *Sophist* 242e–243a, where Empedocles is explicitly evoked), the "harmony between human and beasts, without warfare or discord" (CK 150–151). Let me note in passing that Kahn is, in my view, absolutely right to speak here of *two* opposed situations. The *tote men . . . tote de* construction is just one of the most visible *grammatical* structures marking this duality. The thematic and conceptual structures supported by this grammar end up informing not only the entire myth but the entire dialogue. I thus find utterly unconvincing the three-stage theses of either Luc Brisson or C. J. Rowe in *Reading the Statesman*, ed. C. J. Rowe (Sankt Augustin, Academia Verlag, 1995). See also Melissa Lane's treatment of this question in *Method and Politics in Plato's "Statesman"*, 103–104.

32. Miller makes a similar point: "The principle is the analogy of micro- to macrocosm: human existence 'follows and imitates' the life of the cosmos (274a)" (MM 39).

33. We might speculate, in light of the *Statesman* and its divine pilot who at times steers the universe and at other times withdraws, that Plato is perhaps also rewriting and so reorienting here Heraclitus's famous fragment *ta de panta oiakizei keraunos*: lightening [*keraunos*] steers [*oiakizei*: related to *oiax*] all things [*ta de panta*]. This word *oiax* is the very one used by Plato in the *Statesman* to mark the moment when the God, the pilot or *kybernētēs* of the universe, lets go of the tiller or helm (the *oiax*) and withdraws from the universe at the end of the Age of Kronos in order to let the universe turn backward of its own accord—that is, without his guidance (272e). It is as if Plato in the *Statesman* were rewriting not only Hesiod's myth but Heraclitus's fragment that says it is the thunderbolt, Anaximander's fragment that says it is the *apeiron*, and Epimarchus' fragment that says it is the Logos, that "steers all things." Where these other thinkers locate the thunderbolt, the

apeiron, or the Logos, Plato finds a God or Demiurge endowed with intelligence—a God who is no longer one of the many things in the universe but their overseer, their pilot, even their father and their teacher, a father and a teacher to be imitated and learned from in the Age of Zeus.

3. THE SHEPHERD AND THE WEAVER: A FOUCAULDIAN FABLE

1. Michel Foucault, *Security, Territory, Population: Lectures at the Collège de France, 1977–1978*, trans. Graham Burchell (New York: Picador, 2007); *Sécurité, Territoire, Population, Cours au Collège de France, 1977–1978* (Paris: Éditions du Seuil/Gallimard, 2004). Hereafter abbreviated *STP*, with pages references to the English edition and then the French.

2. As Miller claims, "by his account of the god's relation to the cosmos, the Stranger provides a transcendent measure for statesmanship" (*The Philosopher in Plato's "Statesman"*, together with "Dialectical Education and Unwritten Teachings in Plato's *Statesman*" [Las Vegas: Parmenides Publishing, 2004], 37; hereafter abbreviated MM). The Age of Kronos thus remains, even if transcendent, a measure for both the human statesman and the cosmos itself in the Age of Zeus. "In its form of motion," for example, "the cosmos imperfectly imitates the god. . . . [even though] 'nothing except' the god himself (269e) may have such perfect motion" (MM 38).

3. Miller wavers a bit on this shift in figures, though on balance he seems to support the view that Plato never completely replaces or rejects the figure of the statesman as shepherd but instead integrates it into a higher order. Miller begins by asking, like Foucault, why Plato would "put forth and then reject this figure [of the shepherd], replacing it with that of the weaver" (MM xiv) and he argues that "the initial *diairesis* and the myth, as position and refutation, constitute a clear rejection of the reemergence of despotism in Greek politics" (MM 54). But his subsequent analyses do not support such a rejection. He later writes, for example, that "if the contraposition of the two ages is the Stranger's means of showing what man *cannot be* in the Age of Zeus, it is also his means of showing, positively, what the god *is* in the Age of Kronos. And this vision may serve as a 'measure,' if not of man's actual possibilities, at least of what, through these possibilities, he may strive towards" (MM 51). This helps explain Miller's frequent uses of the term *analogue* throughout his analysis of the *Statesman*. Miller speaks, for example, of statesmanship in the Age of Zeus as "distinct from the Kronian shepherd," and yet, "as a 'care for the whole' which seeks, within the age of Zeus, to reverse its increasing disorder or 'unlikeness,' it is the proper human *analogue*, the appropriation, of the god's rule" (MM 52–53; my emphasis). Elsewhere, he argues that "the arts are the proper *analogue* in the Age of Zeus to the gratuitous gifts and nature by the earth and shep-

herd" (MM 52, my emphasis; see MM 60). Indeed, in the Age of Zeus, "man in effect assumes a relation to himself *analogous* to that previously borne to him by the god; as a political animal, he internalizes the god's care, becoming demiurge and harmonizer of his own existence and producing—to the extent that he can 'remember'—a godlike order in the human condition" (MM 77; my emphasis). Much later in his analysis, as he is trying to explain the return of the model of the statesman-shepherd at a point in the dialogue (295e) where this model seems to have been definitively overcome or refuted, Miller is forced to conclude—or to concede—that "the statesman is *like* the ultimate shepherd, the god," and that, "as the unrecognized lawgiver, the statesman is the analogue to the god" (MM 95; Miller's emphasis). Hence Miller ends up arguing that the weaver or statesman in the Age of Zeus is the *analogue* of the shepherd in the Age of Kronos. He therefore recognizes the continuing importance of the myth in the dialogue and, thus, the continuing relevance of the model of the statesman as shepherd. He argues, for example, that Young Socrates "confirms the Stranger's affirmation of the true statesman as 'like a god among men' (303b)—an affirmation which, incidentally, recalls the myth's notion of likeness to the god in setting him above the constraints, including laws, of other polities" (MM 102). Such references to an *analogue* between the Age of Zeus and the Age of Kronos continue right up to the very end of the dialogue, on Miller's account: "With his 'remedy' or 'medicine' (*pharmakon*, 310a) for this sickness, the statesman, in turn, relates to the city analogously as the god relates to the cosmos" (MM 109). On this same point, see David A. White, *Myth, Metaphysics, and Dialectic in Plato's "Statesman"* (Burlington, Vt.: Ashgate Publishing Company, 2007), 126; hereafter abbreviated DW. Finally, Miller not only claims such an analogy between the statesman in the Age of Zeus and the god in the Age of Kronos but provides page references to establish it: "*like* the god, the statesman expresses his 'care for the whole' by 'integrating' (308c, recall 273b) and 'harmonizing' (309c, recall 269d) the opposite factions within the polis" (MM 109; Miller's emphasis). I could not agree more. But because the god is not himself a weaver, the statesman would seem to be imitating, through his weaving, the god as shepherd. The model of the statesman as shepherd is, therefore, not simply rejected or refuted but transformed, internalized.

Stanley Rosen, for his part, argues that "the myth is not about two distinct races, but rather about two aspects of human existence" (*Plato's "Statesman": The Web of Politics* [New Haven: Yale University Press, 1995], 44; hereafter abbreviated SR). "Zeus and Kronos," he writes, "are coordinate and reciprocal expressions of the two dimensions of human existence" (SR 53). In the end, Rosen claims, "the Stranger's political doctrine is scarcely different from the views of Socrates in *Republic* or the Athenian Stranger in *Laws*. The human animal cannot finally escape its herdlike nature but must be guided by philosophical shepherds or their surrogates" (SR 7). While Rosen thus echoes on occasion Foucault's claim that "the purely biological standpoint of the original *diairesis* has been entirely replaced by a properly political standpoint" (SR 98), his ultimate position seems to be that the image of the statesman as shepherd is never entirely overcome: "The importance of breeding . . . shows that shepherds or grooms will be required for the human beings in their identity as herd or pack animals; in this sense, the discussion of the nurturing of herds in the *diairesis* was not entirely off the mark" (SR 189).

4. Mika Ojakangas argues along very similar lines in his recent book *On the Greek Origins of Biopolitics* (New York: Routledge, 2016). Ojakangas claims that the intellectual background for pastoral power is to be found not in the Christian pastorate but in classical political thought, and so already in Plato. While Ojakangas focuses his analysis on the *Republic* and *Laws*, devoting an entire chapter to each (59–76, 77–100), the *Statesman* is treated in several places along the way (see 6–7 and, especially, 81–85). Ojakangas's analysis of the importance of the model of the statesman as shepherd accords very much with my own here. The larger claim of Ojakangas's very compelling work is that biopolitics is not, as Foucault would have it, "an exclusively modern idea" but one that can be found already in Greek antiquity. As Ojakangas puts it on the opening page of his work, "The idea of politics as control and regulation of the living in the name of the security, well-being, and happiness of the state and its inhabitants is as old as Western politics itself, originating in classical Greece. Greek political thought . . . is biopolitical to the bone" (1).

5. See *STP* 123/127–128 and 164/167–168 for other formulations of Foucault's thesis.

6. Miller follows the way in which "the notion of the ruler as shepherd to man appears indirectly in the Homeric epics," that is, as the trace of "an earlier notion of the king—a notion which was fully effective, perhaps, only in the pre-Homeric Mycenaean period" (MM 40). But he then shows that "there arose a new interest in kingship during the fourth century" (MM 46) and a return to the archaic figure of the statesman as shepherd. He shows how, from Xenophon's *Cyropaideia* through Isocrates, there was, as a response to the "rise of sophistic humanism" (MM 44), renewed interest in this "vision of a natural strong man, one who can free himself of the constraints imposed by democracy," in short, "a resurfacing of the despotic current in Greek political history" (MM 46). This allows Miller to interpret the Platonic references to the statesman as shepherd (*Republic* 345b–d) in a much more positive light and to conclude that "Socrates's use of the

shepherd-image seems substantive" (MM 48), so long, that is, as the shepherd-king is willing to "become genuinely philosophical" (MM 48). As Miller asserts, "Hesiod's description of the paradisiac life under Kronos evidently answered to a deep, persevering passion in the Greek soul" (MM 43). Hence "'the beginning' stage in the Age of Zeus refers to the period of ancient despotism which is glorified, even in its demise, by Homer and Hesiod, the period when godlike kings, literal copies of the shepherd-god, ruled absolutely as 'shepherds of the people'" (MM 49).

This fits well with Miller's overall interpretation of Plato's political thought. Miller speculates at the end of his analysis that "Plato's figure of the weaver-statesman, designed to orient the law-state, is . . . an 'interim device' or stop gap," like Penelope's weaving. For Plato perhaps hoped that, one day, a philosopher-king might reveal himself, like Odysseus returned to Ithaca, and that this would constitute "Plato's true return and restoration" (MM 118). "In making his trip to Syracuse in 367 Plato seemed to hope for just this" (MM 48). Miller thus places the writing of the *Statesman* around 362 or later—well after the disappointment of Syracuse (MM 116–117).

7. Foucault refers to *Critias* 109b–c; *Republic* I, 343a–345e; III, 416a–b; IV, 440d; *Laws* V, 735b–e; and *Statesman* 267c–277d.

8. Miller argues for something similar when he suggests that the "overcoming of humanistic 'forgetfulness' requires a new form of memory," a kind of "remembrance" that "transcends Hesiod as well as Protagoras; as the recovery of rational principle, it is *anamnēsis*, or philosophical recollection" (MM 51).

9. David White has shown, perhaps better than anyone, just how the myth of the two ages is "the pivot of the entire dialogue" (DW 1; see 6). As White shows, the myth presents two extremes, one of excess (the Age of Kronos) and one of deficiency (the Age of Zeus), which are woven together in a way that resembles the very activity of the statesman. The statesman is modeled not just on the shepherd in the Age of Kronos ("The true statesman resembles the divine shepherd identified by the myth's complex narrative as the demiurge" [DW 58]) but on the activity of a demiurge who weaves together the two ages. As White contends, "this degree of autonomy [in the cosmos] occupies a *mean between extremes*—between natures fully under control of the demiurge and natures left alone, without divine guidance of any sort" (DW 54). In other words, for White, "this pair of cycles represents a study in excess and deficiency" (DW 56). The myth thus prefigures the definition of the statesman at the end of the dialogue—that is, the conception of the statesman as a weaver of opposing virtues. White's conclusion is illuminating and convincing: "This cyclical sequence of opposites becomes essential in determining the nature of the king. . . . The

result is a delicate harmony of opposites balancing the static perfection of a cosmos defined by unlimited divine beneficence (the era of Kronos) and the chaos of complete cosmic dissolution (the era of Zeus). The human animal oscillates between a cycle when deity bestows everything for our apparent well-being and a subsequent cycle when deity is absent and eventually provides nothing" (DW 57). What the demiurge does on the cosmic level—that is, "weaving together the extremes of two cosmic cycles . . . until, through the process of such weaving, a condition defined by the mean results" (DW 57)—the statesman does on the level of the polis. As a result, the "ruler caring for human beings as if they were a herd of animals must 'weave together' a series of opposites according to the mean, just as the demiurge approximated a mean in preserving the life of human beings within the opposing cycles of the cosmic drama" (DW 58).

10. No less an interpreter of Plato than Charles Kahn argues for the centrality of imitation to the *Statesman*, even if, and he is surely right, this "notion is applied in several different ways, not always consistent with one another" (Charles Kahn, "The Myth of the *Statesman*," in *Plato's Myths*, ed. Catalin Partenie [New York: Cambridge University Press, 2009], 157); hereafter abbreviated CK). Without this notion of imitation, it would be impossible to link the six inferior regimes to the seventh ideal one, or, indeed, the Age of Zeus to the Age of Kronos. Kahn argues that the myth serves as a "device for removing the ideal Statesman from the human world and relocating him in the mythical space of an alternative cosmic cycle. . . . To the extent that the divine ruler of the myth parallels the true Statesman, he remains relevant to the constitutions of our world, since they must imitate the wisdom of his rule as best they can" (CK 160). The fact that, as Kahn puts it, "the true Statesman turns out to be not of this world after all," and that "the seventh form is 'separated from all the other constitutions, like a god separated from human beings' (303b)" (CK 160), does not therefore mean that this true statesman or this seventh regime is not a model to be imitated. On the contrary, it is precisely because the true statesman is elsewhere and the seventh form is separated from the others that they *must* be imitated: "The contrast drawn here between the preferred constitution and 'things as they now are' (*nun de*) means that, in the human world as we know it, there is no true *politeia*, but that we must be satisfied with imitations, some better and some worse, and the better ones all presupposed the rule of law" (CK 159). Of course, Kahn is not the only critic to have argued for the importance of imitation to the *Statesman*. Kahn cites both Grube and Sabine with approval (CK 163), before concluding: "the notion of imitation serves, both in the *Statesman* and in the *Laws*, to bind the second-best solution to the original vision of the *Republic*" (CK 164).

11. We find a similar characterization in the *Laws,* where the Athenian argues that laws are made for the benefit of those ruled and not—as Thrasymachus (*Republic* 338c) or Callicles (*Gorgias* 483d) might claim—for those who rule and who believe that justice is "what benefits the stronger" (*Laws* 714b).

12. This is, perhaps not surprisingly, one of the principal passages from the *Laws* used by Mika Ojakangas to build his case for a biopolitics in Greek antiquity (see *On the Greek Origins of Biopolitics* 8, 93–95).

13. See Vidal-Naquet 377.

14. Charles Kahn comments: "In the myth of the *Laws* as in the doctrine of the *Statesman,* the ideal ruler serves as a model for human imitation, and in both cases the best imitation is a city ruled by law. What is new in this context is that the rule of law is represented as the expression of reason (*nous*). Plato thus seeks to overcome the conflict between law and knowledge that was the center of focus in the *Statesman*" (CK 162).

15. This passage from the *Laws* recalls a passage near the end of the *Statesman* that we will look at in some detail in Chapter 6, the passage in which the human is said to be a combination of the animal and divine and where the work of law involves the binding of these various parts (309c). As Kahn argues in his analysis of the myth, "the notion that the true statesman is more divine than human will turn out to play a constructive role in later sections of the dialogue" (CK 152).

16. On this point, see Vidal-Naquet 371–374.

17. In *Alcibiades I,* for example, Socrates tries to steer the young Alcibiades toward acting out of knowledge rather than ignorance by appealing to Alcibiades's common experience and everyday reliance upon the expertise and knowledge of others. His prime example here is, interestingly, the helmsman. Socrates asks Alcibiades whether, when sailing, he would "think whether the tiller [*oiaka*] should be moved inward or outward, and in your ignorance bewilder yourself, or would you entrust it to the helmsman [*kybernētēi*], and be quiet?" Alcibiades answers: "I would leave it to him," (117c–d). Socrates here guides Alcibiades not to some definite, definable knowledge, but to a recognition of the benefits of knowing what we do not know and entrusting ourselves to the knowledge of others in such cases.

18. Rosen, too, sometimes speaks, like Foucault, of "the rejection of the divine shepherd as a paradigm of the statesman" (SR 102), but he ultimately argues that no one model can really serve as "a model of statesmanship because of the comprehensive nature of the royal art" (SR 106; see 116).

4. THE MEASURE OF LIFE AND LOGOS

1. Rosen makes a similar point when he suggests that "[it is] compatible with the Stranger's procedure for us to infer that *eidē* or common looks can

be constructed as well as discovered" (*Plato's "Statesman": The Web of Politics* [New Haven: Yale University Press, 1995], 17–18; hereafter abbreviated SR).

2. To put it provocatively, man is indeed, for Plato, the measure of all things, man's body the measure of what discourse should look like and man's speech the measure of what speech should sound like. That is why the first and best example is always, in Plato, the example of man. In the *Laws*, for example, the Athenian argues that in order to judge whether some animal has been correctly represented one cannot be "totally ignorant as to what animal was being represented [*to memimēmenon zōon*]?" It is a general claim that is quickly filled in by the Athenian with the example of the face and figure of man: "Well, suppose we should know that the object painted or molded is a man . . ." (*Laws* 668e–669a).

3. See *Meno* 85c on right opinion as "like a dream." As Rosen suggests, "The difference between waking and dreaming is somehow suggestive of the difference between the normal and the counternormal epochs in the Stranger's myth. A myth is more like a dream or prophetic vision than a discursive analysis" (SR 79). To follow the analogy, the dream knowledge of the Age of Kronos must somehow become waking knowledge in the Age of Zeus.

4. Rosen again: "what counts as a model is thus relative to our intentions or, more precisely, to what we intend to understand or explain" (SR 82).

5. The example of letters can be found in several places in the dialogue, from *Republic* 368d to *Theaetetus* 202e–205c and *Sophist* 253a, to name just three. Derrida's "Plato's Pharmacy" is an attempt to think the logic governing all these examples.

6. Plato's theory of language acquisition is thus remarkably similar to someone like Saussure's on this score. Neither think that there are fully formed semantic unities that precede the interplay of similarities and differences, and yet both try to protect speech from writing under the pretext that this latter brings absence, difference, and, ultimately, death along with it.

7. See SR 98–118 and Seth Benardete's translation of *Statesman* (Chicago: University of Chicago Press, 1986), 107–113. David White argues: "The Stranger's paradigmatic account of the paradigm of wool weaving also reveals the implicit presence of the myth. If composition and separation are found in 'all things' and this includes the cosmos itself as well as everything in the cosmos, then composition and separation pertain not only to the production of the cosmos as a uniquely ordered and living whole but also to every ordered and living whole existing within the cosmos. The myth speaks of the demiurge as a 'harmonizer' (269d) and, later, as a 'composer' [*synthentos*—273b]" (*Myth, Metaphysics, and Dialectic in Plato's "Statesman"*

[Burlington, Vt.: Ashgate Publishing Company, 2007], 73; hereafter abbreviated DW).

8. Here is a very brief summary of this *diairesis* of weaving, which is itself, as Benardete and others have noted (see DW 75), the most complete account of ancient Greek weaving we have: All things we make or acquire can be divided into those with which we do something and those for defense against suffering (hence the opposition here is between active doing and passive suffering); in this latter category, there are non-material things (spells and antidotes) and material defenses; of these latter, some are equipment for war and some are protections; of these latter, some are screens and some are defenses against heat and cold; of defenses, some are shelters and others coverings; among coverings, there are rugs that go under us and things that wrap around us; of these latter, some are made of one piece and others of several pieces; of these latter, some are stitched and some are not; of the unstitched, some are made of plants and some of hair; of those made of hair, some are cemented together and some are fastened without any external adhesive; it is to these latter "protective coverings made of materials fastened without extraneous matter" that we give the name "clothes" (*himatia*), and to the art that corresponds to the production of these things the name "clothes-making" (*himatiourgikēn*) (280a). Having thereby isolated or separated off the art of clothes-making from its kindred arts, the Stranger goes on to argue that they have really isolated and defined the art of weaving as well, since the greatest part of weaving concerns clothes-making and so differs from it, he suggests, only in name, just as the royal art (*basilikē*) and statecraft (*politikē*) were earlier shown to differ in name only. After distinguished weaving from its "*kindred* arts [*xyngenōn*]," the Stranger goes on to distinguish it from "closely *co-operative* arts [*xynergōn*]," that is, from arts such as carding, mending, washing, and caring for clothes, and then the making of tools for weaving, spindles and shuttles, the warp and woof, and so on (281c–282b).

9. Hence we "shift from war to peace in order to arrive at weaving, which is not only a defensive art but also one that is practiced in the house by women" (SR 103; see 153).

10. White calls these two forms of measure "quantitative" and "mean (or due)" measure (DW 12). Rosen calls them "arithmetical and nonarithmetical," and he compares them to Pascal's *esprit géométrique* and *esprit de finesse* (SR vii). With the former, there are just two relevant terms, "correct and incorrect," while with the latter there are "excess, deficiency, and suitability" (SR 120). Rosen later writes: "Apparently the Stranger regards the doctrine of the mean to be his own discovery; one should note that Aristotle's innovation amounts to a narrowing of the domain of this doctrine to ethical

virtue" (SR 134). Miller speaks of "relative" measure as opposed to "essential" measure"; accordingly, "the mean serves as the norm for *praxis*, the standard by which essential measure can judge speeches and actions" (Mitchell Miller, *The Philosopher in Plato's "Statesman"*, together with "Dialectical Education and Unwritten Teachings in Plato's *Statesman*" [Las Vegas: Parmenides Publishing, 2004], 66–67).

11. David White gives this contemporary (quintessentially American) example: "A 12-inch piece of wood is too long (excessive) for a toothpick and too short (deficient) for a baseball bat" (DW 82).

12. Rosen makes a similar point: "In nonarithmetical measurement, everything measured is excessive, deficient, or suitable with respect to a human purpose. Nothing is excessive, deficient, or suitable in itself. Every measurement of this sort is relative to a frame of reference" (SR 121).

13. As Rosen argues, "the fundamental question raised by the *Statesman* is then not at all that of the nature of the royal art, but rather the nature of dialectic" (SR 100).

5. FRUITS OF THE POISONOUS TREE: PLATO AND ALCIDAMAS ON THE EVILS OF WRITING

1. "Plato's Pharmacy," in *Dissemination*, trans. Barbara Johnson (Chicago: University of Chicago Press, 1981), 63; *La dissémination* (Paris: Éditions du Seuil, 1972), 71; hereafter abbreviated "PP" with pages references to the English edition and then the French. Just a couple of pages later, Derrida will speak of all the "paradoxes of supplementarity, and the graphic relations between the living and the dead: within the textual, the textile, and the histological" ("PP" 65/73).

2. Derrida cites *Statesman* 277d–e in a footnote and then promises to return to that dialogue at the end of his essay. This is the itinerary that I, too, will follow in order to illuminate both the theme of life and the subtle but important role played by Alcidamas in the *Statesman*.

3. While Theuth, the servant and son, thus presents writing as a pharmakon or *remedy* for memory, Thamus, the master and father, judges it to be pharmakon or *remedy* only for reminding. This evaluation lets it be understood that, as a pharmakon, writing is a *poison* for living memory.

4. See my essay "Earmarks: Derrida's Reinvention of Philosophical Writing in 'Plato's Pharmacy,'" in *Derrida and Antiquity*, ed. Miriam Leonard (New York: Oxford University Press, 2010), 43–72.

5. Because life is now no longer simply *opposed* to death, Derrida speaks of writing in terms of *ghosts, fantômes*, which "can no longer be distinguished, with the same assurance, from truth, reality, living flesh, etc." ("PP" 104). In the end, "writing is not an independent order of significa-

tion; it is weakened speech, something not completely dead: a living-dead, a reprieved corpse, a deferred life, a semblance of breath" ("PP" 143), "half-dead discourses . . . persecuted for lack of the dead father's voice" ("PP" 146).

6. In *Euthydemus*, Socrates suggests that only things that "have life" (*ta psychēn echonta*) "intend" (*noei*) and that phrases, *rhēmata*, which are lifeless (*ta apsycha*), do not intend (*Euthydemus* 287e).

7. Derrida, already in 1968, speaks of *khōra* in this regard: "the problematic of the *moving* cause and the *place*—the third irreducible class—the duality of paradigms (49a), all these things 'require' (49a) that we define the origin of the world as a *trace*, that is, a receptacle. It is a matrix, womb, or receptacle that is never and nowhere offered up in the form of presence, or in the presence of the form, since both of these already presupposed an inscription within the mother" ("PP" 159–160). Derrida cites a long passage from *Timaeus* and then writes, "The *khōra* is big with everything that is disseminated here. We will go into that elsewhere." ("PP" 160–161) These allusions to fecundity and dissemination already anticipate, some two decades before the fact, that the essay "Khōra" (1987), will also revolve around the question of life.

8. Derrida writes: "The association of logos-zōon appears in the discourse of Isocrates *Against the Sophists* and in that of Alcidamas *On the Sophists*" ("PP" 79n12).

9. Alcidamas, *On the Sophists*, translated by LaRue Van Hook (Classical Weekly, January 20, 1919); hereafter abbreviated *OS*. I have also consulted and benefited from the translation of J. V. Muir in *Alcidamas: The Works & Fragments* (London: Bristol Classical Press, 2001). I would like to thank Marina McCoy of Boston College for reminding me of the many similarities between Alcidamas's critique of writing and Plato's.

10. Alcidamas will return to this point at the very end of his discourse, arguing that it is not at all "illogical for one to condemn written discourse" by means of written discourse (*OS* 29). For his goal is not "altogether to condemn the ability the write," only to show that it is "of lesser worth than extemporaneous speaking" and that one should "bestow the greatest pains upon the practice of speaking" (*OS* 30). In short, it is no contradiction to argue for both the inferiority of writing and its usefulness for some purposes.

11. Writing, he says, weakens the intellect of the writer himself and makes him unable to speak effectively: "the practice of writing, by making sluggish the mental processes, and by giving the opposite sort of training in speaking, produces an unready and fettered speaker, deficient in all extemporaneous fluency" (*OS* 17).

12. For the relationship between Plato's views on writing and speaking and Gorgias's, and in particular with regard to the *kairos*, see Melissa Lane, *Method and Politics in Plato's "Statesman"* (New York: Cambridge University Press, 1998), 150–152.

13. A further advantage is to be found in the fact that "in extemporaneous speaking forgetting involves no disgrace," for one can then simply move on to the next point and no one will really notice (*OS* 20). Such lapses in a written discourse are much less easy to conceal, "since, if the slightest detail be omitted or spoken out of place, perturbation, confusion, and a search for the lost word inevitably follow, and there ensues loss of time—sometimes, indeed, abrupt silence and infelicitous, ludicrous and irremediable embarrassment" (*OS* 21).

14. Isocrates, *Against the Sophists*, in *Isocrates II*, trans. George Norlin (Cambridge: Harvard University Press, 1962). In this speech, Isocrates criticizes those who do not recognize that "the art of using letters remains fixed and unchanged, so that that we continually and invariably use the same letters for the same purposes, while exactly the opposite is true of the art of discourse" (12). For "oratory is good," he goes on to say, "only if it has the qualities of fitness for the occasion [*kairōn*], propriety of style [*prepontōs*], and originality of treatment, while in the case of letters there is no such need whatsoever" (13). As Isocrates will go on to argue, the difficulty of speech has to do not with gathering together the various elements for discourse but with choosing and arranging them "properly" so as not "to miss what the occasion demands [*kairōn*]" and adorning them "in an appropriate way [*prepontōn*]" (16–17; translation slightly modified).

15. This is one of just two uses of the term *kairos* in the entire dialogue. The other is at 229a, where Phaedrus says he is fortunate [*etuchon*] to have found himself, for this occasion [*eis kairon*], barefoot, while Socrates always is (*Phaedrus* 229a).

16. Stanley Rosen echoes many of the sentiments of Alcidamas: "There is a fundamental difference between speaking and writing. In the case of speaking, we can redesign the length or structure of the speech as we talk, in response to our perception of the reactions of the audience. This is not possible in a writing which, however cunningly constructed, can only take into account the varying natures and circumstances of its potential readers up to a point and in an approximate manner" (Rosen, *Plato's "Statesman": The Web of Politics* [New Haven: Yale University Press, 1995], 122; hereafter abbreviated SR). Rosen identifies this capacity for adjusting speech to one's audience with *phronēsis*: "One thinks here of Socrates' criticism of writing in the *Phaedrus*. Speech is superior to writing because it can accommodate its assertions to the nature of the immediate audience. The voice of *phronēsis* is

capable of innovation; writing is always the same, and *nomos* is the least mobile of writings" (SR 158).

17. Rosen opposes politics to the true art of statesmanship, which he calls philosophy: "Politics is oriented toward the body; but philosophy, or the genuine art of statesmanship, is oriented toward the soul" (SR 162).

18. Phaedrus responds to Socrates's claim in this way, "O Prodicus! How clever [*sophōtata ge*]" (*Phaedrus* 267b). As is well known, Prodicus plays a curiously singular role in Plato's dialogues; he is at once poked fun at and, it seems, held in some esteem. Just a few lines later, Socrates gets Phaedrus to agree that when it comes to writing tragedies and comedies one must know not just how to write long and short speeches but how to "properly" (*prepousan*) combine these into a whole (*Phaedrus* 268d).

19. The measure of the mean will be essential to the physician or the trainer, for example, who, when alive and present, will alter his prescriptions or his regimen and tailor them to the specific needs of the individual rather than offer a general prescription to suit the majority.

20. It is found twice in the *Euthyphro* (5a, 16a), twice in the *Menexenus*, and then just once in *Apology* (20a), *Cratylus* (413d), *Euthydemus* (278e), and, significantly, *Phaedrus*.

21. I am speaking here of Plato's *use* of the term *autoschediazein* and not of Socrates's actual speech in the *Menexenus*. For an excellent analysis of this latter, see Nickolas Pappas and Mark Zelcer, *Politics and Philosophy in Plato's Menexenus* (New York: Routledge, 2014).

22. "What Plato is attacking in sophistics," writes Derrida, "is not simply recourse to memory but, within such recourse, the substitution of mnemonic device for live memory, of the prosthesis for the organ . . . the passive, mechanical 'by-heart' for the active reanimation of knowledge, for its reproduction in the present" ("PP" 108).

23. As Derrida suggests, "The argumentation against writing in the *Phaedrus* is able to borrow all its resources from Isocrates or Alcidamas at the moment it turns their own weapons . . . against the sophists. Plato imitates the imitators in order to restore the truth of what they imitate: namely, truth itself" ("PP" 112).

24. Derrida in *Glas* says this, for example, of Hegel's treatment of *life* in the greater *Logic*: "In the last section of the 'Subjective Logic' ('The Idea'), life is inscribed both as a syllogism and as the moment of a syllogism. . . . In this syllogism of the Idea, life first appears as a natural and immediate determination: the spirit outside self, lost in naturality, in natural life that itself constitutes a 'smaller' syllogism. The immediate Idea has the form of life. But the absolute Idea in its infinite truth is still determined as Life, true life, absolute life, life without death, imperishable life, the life of truth."

Glas, trans. John P. Leavey, Jr., and Richard Rand (Lincoln: University of Nebraska Press, 1986), 82a.

25. It is also the dream, in short, of an "immediate sign." See Jacques Derrida, *Of Grammatology*, trans. and preface by Gayatri Chakravorty Spivak (Baltimore: Johns Hopkins University Press, 1976), 233–235.

26. Rosen writes, "the art of politics, understood as commanding, would have to be supplemented by the art of rhetoric and not by the art of pure philosophical rhetoric alone" (SR 152).

6. THE LIFE OF LAW AND THE LAW OF LIFE

1. For example, says the Stranger, the group of "coins, seals, and stamps can be forced" into the class of "ornaments" or else "instruments" (289b).

2. These seven classes contain everything except tame animals, which the Stranger then goes on to enumerate. Miller has a detailed account of why the Stranger shifts from bifurcatory diairesis, the only kind of diairesis he uses in the *Sophist* and the kind he begins with in the *Statesman*, to this non-bifurcatory kind (see Mitchell Miller, *The Philosopher in Plato's "Statesman"*, together with "Dialectical Education and Unwritten Teachings in Plato's *Statesman*" (Las Vegas: Parmenides Publishing, 2004), 73–113, especially 73–86, and then 142–157??; hereafter abbreviated MM). While there is much that "points to the crucial role of bifurcatory method" (MM 26), Miller argues that there are "difficulties, if not deficiencies, with bifurcatory diairesis" (MM 27). For Miller, the move to the non-bifurcatory shows the possibility of "letting the eidetic structure of things themselves over-ride [the rules of bifurcation]" (MM xvi). As he puts it later, it is intended to let the "working community [of the things themselves] appear" (MM 77), and this, he suggests in an ingenious supplement to his original work, is in conformity with the "'unwritten teaching' that Aristotle credits to Plato in *Metaphysics* A6" (MM xvii). Miller argues, in effect, that what we have taken to be a seven-kind division is, upon closer analysis, a fifteen-kind division—a fifteenification, as he calls it—corresponding, in Greek musical theory, to the double octave. Hence, we have among the tame animals, a category that has already been divided so as to separate off all non-human animals, slaves (an eighth class), merchants (ninth), heralds and clerks (tenth)—none of whom are genuine pretenders to the art of statesmanship—but then interpreters, priests, and prophets (eleventh), the orator (twelfth), the general (thirteenth), the judge (fourteenth), and finally the statesman (fifteenth) (see MM xvii, 150–157).

3. Hence the name democracy is given to *both* the regime in which the multitude rules with laws and with the consent of the ruled *and* the regime in which the multitude or the mob rules without laws and without the consent of the minority.

4. The more mild kind, the sort to be carried out during a time of scarce resources, consists in exiling a segment of the population and then euphemistically calling this an "emigration." Fortunately for the Athenian and his Cretan and Spartan counterparts, who are contemplating the best way to settle a *new* and not yet existing city, they do not need to purge or purify their city. They must instead be careful in their initial selection of citizens, drawing off certain streams or sources that might "pollute" the pool in order to keep the city as clear and pure as possible.

5. In the *Laws*, it is not the biological father but, interestingly, the officer appointed by the *polis* to preside over the entire educational program of boys and girls who is characterized as their "legitimate father" (*gnēsiōn patēr*) (*Laws* 765d). Of all the offices of the state, this one is most important, says the Athenian, for the "first shoot" is most important; if it sprouts well it is "most effective in bringing to its proper development the essential excellence of the creature in question" (*Laws* 765e).

6. For Stanley Rosen, "the central theme of the *Statesman* is the relation between *phronēsis*, or sound judgment, and *technē*" (Stanley Rosen, *Plato's "Statesman": The Web of Politics* [New Haven: Yale University Press, 1995], vii; hereafter abbreviated SR). The whole question, argues Rosen, is one of knowing whether there can be a "rule of *phronēsis* unencumbered by *nomos*," which, for Rosen, is the product of *technē*. Rosen's answer to this question is, quite simply, *no* (SR viii). While Plato defines statesmanship throughout the dialogue in terms of various forms of *technē*, such as weaving, the genuine statesman is someone who exercises *phronēsis*—that is, a form of rule without *nomos* or *technē*. Rosen writes: "*Phronēsis* and *technē* are entirely distinct from one another; hence too *phronēsis* is distinct from *nomos*, which is a product of *technē*. *Phronēsis* is the genuine king" (SR 162). Indeed, for Rosen, "the genuine statesman" is ultimately "personified as *phronēsis*" (SR 190), and that helps explain why Plato "criticizes the rigidity of law and praises the extreme flexibility of *phronēsis*" (SR 123). But insofar as a genuine rule of *phronēsis* is not realizable in this world, *nomos* and *technē* must be used, on Rosen's account, to compensate for or, as I would say, *supplement* the absence of a genuine statesman. While it is "best if the kingly man were to rule by *phronēsis* rather than that the laws should rule" (SR 156), in the end "*phronēsis* must submit to legislation and thus to the productive art of the statesman" (SR ix).

While Rosen's overall reading of the dialogue is surely correct—a single true or genuine regime or statesman is to be distinguished from various imitations of these—his argument for that reading is misleading. For Plato does not distinguish as clearly as Rosen leads us to believe between *phronēsis* and *technē* (or even *epistēmē*) and Rosen must force his reading of the dialogue to conform to this distinction. (Rosen's insistence upon this distinction is all the more curious in light of his recognition, elsewhere, of the flexibility of Plato's vocabulary: "There is for Plato a master element in the soul, called *nous, dianoia,* and *phronēsis* in different contexts, which ought to rule by nature over the other elements" [SR 69].) Rosen makes it sound as if Plato uses the term *phronēsis* throughout the *Statesman* to describe the activity of the genuine statesman, whereas the term is, in fact, used in its nominal form just twice in the entire dialogue, once to describe the movement of the cosmos in the Age of Zeus (269c) and once—in support of Rosen's argument—as the Stranger says that "the best thing is not that the laws be in power, but that the man who is wise [the man with *phronēsis*] and of a kingly nature be ruler" (294a). In other places, *technē* and *epistēmē* are used to describe the same activity of the true ruler. For example, when Rosen writes, parsing the Stranger's words at 292e, "since *phronēsis* is possessed only by one, two, or a very few, it is necessary to reverse course and write down all the laws of the regime" (SR 174), the term the Stranger uses here is not *phronēsis* but *technē*.

This conflation of terms ends up influencing Rosen's understanding of the relationship between the six imitative regimes and the one true one. Rosen begins by arguing that the Stranger "will identify six kinds of actual regime, all of which are imitations of the seventh, which is alone genuine but which exists only in speech, not deed. . . . Only the genuine regime is ruled by a genuine statesman; the other rulers must be sophists, who were identified yesterday in the conversation with Theaetetus as magicians and masters of disguise" (SR 149). Rosen's appeal to the difference between speech and deed and his allusion to the *Sophist* are helpful, but Rosen goes on to complicate the issue when he suggests that "the Stranger establishes as his seventh type what [he] will call the epistemic city, of which all the other regimes are better or worse imitations (297c)" (SR 167). Instead of speaking, as the Stranger consistently does, of six regimes that imitate the one true regime of *phronēsis/technē/epistēmē*—that is, a seventh regime in which the ruler "makes his science [his *technē*] his law" (297a)—Rosen argues that the six regimes imitate an *epistemic* regime, which is itself an imitation of the regime of *phronēsis* ("the epistemic statesman," he writes, "is himself an image of the original of the rule of *phronēsis*" [SR 179]). We thus end up with eight regimes, not seven.

It's an interesting thesis, and one can understand why Rosen would want to make it, but it is not supported by the dialogue. In the end, it is "overingenious" (*SR* 100)—to use Rosen's own way of characterizing the way others tend to view his work. Rosen is right to suggest that Plato in the *Statesman* is rethinking the role and nature of *epistēmē* and *technē* in statesmanship and that the dialogue complicates the picture we get in the *Republic*.

But there is no simple opposition between rule by *phronēsis* and rule by *epistēmē*, as Rosen, no doubt influenced by his reading of Aristotle, suggests (see SR 127).

7. Miller puts this very well: "The stranger accepts law but only as instrumental for epistemic statesmanship. This directly opposes Young Socrates's objection to 'rule even without laws' (293e), for it puts the true statesman himself above the law" (MM 92–93). "Epistemic statesmanship, then, transcends laws. Whereas it may express itself through them, it cannot be bound to them—not, that is, without ceasing to be *epistēmē*" (MM 93).

8. As Rosen notes, this is also the view of the Athenian Stranger in the *Laws*; "*phronēsis* is the highest of divine goods (631c), but the human possessor of *phronēsis* must rule in accord with law (690b–c)" (SR 158).

9. To illustrate the difference between simply stating the law without preface and prefacing the law with a kind of persuasive prelude—that is, to illustrate the difference between what will be called the unmixed method of law (having one's law straight) and the mixed method (having it with a splash of persuasion)—the Athenian uses the analogy of two different methods of doctoring and, thus, two different kinds of doctor, one who does not look into the causes of disease and does not talk to his patient, who merely prescribes autocratically, without the consent of his patient, and one who talks to his patient, who looks into the origins and natural development of diseases, and who tries to *persuade* his patient to undergo treatment (*Laws* 720c–e).

10. David White argues that "the true statesman imitates in practicing statecraft what the demiurge has achieved within the cosmos as a whole by blending opposites with respect to humanity as the primary resident of the cosmos" (*Myth, Metaphysics, and Dialectic in Plato's "Statesman"* [Burlington, Vt.: Ashgate Publishing Company, 2007], 6; hereafter abbreviated DW). Put otherwise, "an isomorphism emerges between how the demiurge fashioned and preserved the elements of the cosmos and the Stranger's later theoretical formation of measurement" (DW 95).

11. *zōgraphia* is the art of painting or depicting something in a visual medium (see *Republic* 373a, *Sophist* 236b) and a *zōgraphos* is a painter (see *Protagoras* 312c–d; *Gorgias* 454a, 503e; *Republic* 472d, 500e, 501c, 596e, 597b, 598b, 605a; *Theaetetus* 145a; *Cratylus* 424d; *Philebus* 39b; *Laws* 656e, 769a–c, 934c, 956b; *Theages* 126e). The result of this art or this activity of painting—that is, of *zōgraphō* (*Philebus* 40a–b, *Letter VII* 342c)—is a *zōgraphēma*, a picture, an imitation in pictorial form. As the *Cratylus* puts it, "a name is an imitation [*mimēma*], just as a picture is [*hōsper to zōgraphēma*]" (*Cratylus* 430e), and "paintings also are imitations, though in a different way, of things [*ta zōgraphēmata . . . mimēmata einai pragmatōn tinōn*]" (*Cratylus*

430b; see also 430d, 431c, 434a–b, *Hippias Major* 298a, *Philebus* 39d–e). The resulting imitation or picture, a *zōgraphēma*, can even be of an animal—that is, of a *zōon* (*Gorgias* 453c–d, *Philebus* 51c). But even more importantly, the image or picture can itself be called a *zōon* (*Statesman* 277c; *Laws* 769a–b; *Cratylus* 425a, 429a, 430d; and, perhaps, *Republic* 420c), which helps explain the name *zōgraphia*, the art of painting in general, and not just the painting of *zōa*. It is this ambiguity that Socrates plays on at the beginning of the *Timaeus* when he says that he would like to see the *zōon* (the animal in repose) that they have just created in the *Republic* in motion—that is, the *zōon* (the image) that has just been sketched out come to life (*Timaeus* 19b–c).

12. For this image of the ruler as captain of a ship, see *Republic* 488a and *Laws* 963b.

13. Moses Mendelssohn famously described Judaism or Judaic law as a "living script": "The ceremonial law itself is a kind of living script, rousing the mind and heart, full of meaning, never ceasing to inspire contemplation and to provide the occasion and opportunity for oral instruction." *Jerusalem*, trans. Allan Arkush (Hanover and London: University of New England Press, 1983), 102–103. At the 2012 Collegium Phaenomenologicum at which an early draft of this work was presented, there was a live violin performance by Peter Hanly of an original composition by Andrew Lovett entitled, precisely, "A Living Script."

14. But then this problem arises, as Miller well recognizes: "How can one 'imitate' what one does not know?" (MM 103). Similarly, how are the many to recognize the few who know?

15. Rosen writes, suggestively: "In keeping with the *Sophist*, we could also say that the laws are fantasms of *phronēsis*, or false images artfully constructed in such a way as to look like the original to human vision" (SR 158; see 149).

16. Charles Kahn argues that the rule of law is indeed a sort of "second best" (to the regime in which the true statesman makes his science his law) and that, in this regard, the *Statesman* is a sort of hinge dialogue between the *Republic* and the *Laws*. Kahn writes, "Law is the *deuteros plous* (300c; cf. 297e) . . . The city of the *Laws*, in which the rulers must be servants of the law, is repeatedly described as second best (739a, 739e, 807b, 875d)" (Charles Kahn, "The Myth of the *Statesman*," in *Plato's Myths*, ed. Catalin Partenie, 160 [New York: Cambridge University Press, 2009]; hereafter abbreviated CK). "In this regard the *Statesman* and the *Laws* are mirror-images of one another. The former argues fiercely for the supremacy of knowledge, and accepts legality only with regret; the latter trumpets the rule of law and mentions the preference for knowledge only as an aside. . . . The *Statesman* is rightly seen as a transitional dialogue, in which Plato is moving from the

position of the *Republic* to the position of the *Laws*" (CK 161). And yet, as Kahn reminds us, after making reference to the *Crito* and Socrates's decision to obey the laws of Athens rather than escape from prison, "the concern for the rule of law, which is so prominent in these two late dialogues [that is, *Statesman* and *Laws*], is not a new theme in Plato's work" (CK 162–163).

17. For the physician as the *one* "who is worth as much as *many* others," see *Iliad* 12.514.

18. This emphasis on experience is telling. One might wish to compare this to the difference articulated in the *Laws* between the physician who, as it were, makes his science his law by using reason and persuasion to treat free men and the physician, the so-called slave doctor, who simply follows prescriptions based on experience (*Laws* 720c–e).

19. David White's comments on this passage are helpful and underscore the relation between names and imitation. "The Stranger wants one name—'king'—to cover five distinct types of government. Why? Because imposing a single name forces the user of language and the student of statecraft to appreciate that singularity of name reflects singularity of structure, that only *one* true Form of statecraft exists. As a result, the various types of government which have evolved over time and for various reasons must all be understood as deriving from, or imitating, that one true Form" (DW 113).

20. Derrida prefaces his 1967 essay "Form and Meaning: A Note on the Phenomenology of Language" with this epigraph from Plotinus: *to gar ikhnos tou amorphous morphē*, "form is the trace of the formless." The epigraph suggests that Plotinus's notion of the trace was something of a forerunner of Derrida's. Derrida returns to Plotinus at the end of the essay in a long footnote, which concludes with the words that serve as the epigraph to my next chapter. See Jacques Derrida, *Margins of Philosophy*, trans. Alan Bass (Chicago: University of Chicago Press, 1982), 157, 172n16.

21. Miller argues that this harmonization of courage and *sōphrosunē*—this "eugenic doctrine built on the notion of the two natures as genetic strains"—is essentially Socratic in inspiration (MM 107–108).

22. White writes: "This inappropriateness is *akaira* (307b), privative of *kairon*, one of the modes exemplifying due measure; the description evokes the fact that virtues must be practiced in measured ways at certain times— if not, then what was measured and virtuous devolves into its opposite" (DW 123).

23. As Miller points out, this reference to disease underscores the way in which the polis, too, as is often the case in Plato, is like an organism: "As an organic whole, the polis has psyche as well as body; to work together, its 'limbs' require not only interconnections but, too, the life-force which

activates them" (MM 105). This reference to sickness in cities (307d) also recalls the sickness and degeneration of the cosmos in the Age of Zeus (273d–274a, d) (MM 109).

24. The description of weaving in the *Laws* conforms to this description: "the stuff of the warp must be of better quality—for it is strong and is made firm from its twistings, whereas the woof is softer and shows a due degree of flexibility—from this we may see that in some such way we must mark out those who are to hold high offices in the State and those who are to hold low offices, after applying in each case an adequate educational test" (*Laws* 734e–735a).

25. Indeed, as White points out, this final definition of the statesman is not so far off from the results of the initial *diairesis*: "the true statesman characterized at the conclusion of the dialogue closely resembles the divine ruler described in the initial division of the dialogue—both are intellectual (258d–e), directive (260b), and originary (260e). Thus the final version of statecraft appears to run full circle back to the beginning of inquiry into the nature of statecraft" (DW 6).

26. White reminds us: "The 'care' [*epimeleian*—310c] stipulated by the Stranger as the responsibility of prospective parents mirrors, in a more limited social and familial setting, the care [*epimeleian*—cf. 273a, 274b, 274d] which the demiurge directs toward the needs of the cosmos and its inhabitants" (DW 127).

27. For Rosen, "the concluding thesis that the city must be woven together from the two opposing natures of courageous and temperate human beings . . . is anticipated by Socrates in the *Republic* (503b7–504a1), where he indicates the need to mix together the quick and the steady in the souls of the guardians of the just city" (SR 2; see also 185, 189).

28. This activity thus aims for a sort of *mean* between opposing virtues. For Miller, then, the statesman, the person, is the one who realizes the mean (MM 113), while the *Statesman*, the dialogue, "represents a 'mean,' the fullest realization of true polity which is actually possible for the nonphilosophical 'many'" (MM 115).

7. PLATO AND THE INVENTION OF LIFE ITSELF

1. Martin Heidegger, *An Introduction to Metaphysics*, trans. Ralph Manheim (New Haven: Yale University Press, 1959), 71.

2. Plotinus, *Enneads*, trans. A. H. Armstrong (Cambridge: Harvard University Press, 1967), III.6.6.10–15.

3. Giorgio Agamben, *Homo Sacer: Sovereign Power and Bare Life*, trans. Daniel Heller-Roazen (Stanford: Stanford University Press, 1998), 1. See Derrida's critique of this distinction between *bios* and *zōē* in *The Beast and the*

Sovereign, Volume 1, trans. Geoffrey Bennington (Chicago: University of Chicago Press, 2009), 315–316.

4. This phrase can be illuminated, perhaps, by a passage in the *Statesman* where *bios* and *biōtos* are used in a similar way. Young Socrates there affirms, at the Stranger's prodding, that if forced to follow written law rather than knowledge, "all the arts would be utterly ruined . . . and so life [*ho bios*], which is hard enough now, would then become absolutely unendurable [*abiōtos*]" (*Statesman* 299e). In other words, a life, a *bios*, in which the arts must follow written laws rather than living knowledge would be *abiōtos*— that is, unlivable or unworthy of being lived since it and the arts within it would go *unexamined* and so would never be improved.

5. In the *Laws* there is a long passage contrasting a healthy life with a diseased one, a pleasant life with a painful one, a licentious life with a temperate one, a brave life with a cowardly one, a life of bodily and spiritual virtue with a life of vice (*Laws* 734c), and, earlier in the same dialogue, a "life that is just and holy" with an "unjust life" (*Laws* 663d; see also *Republic* 538e, *Phaedo* 114e, *Timaeus* 42b). In the *Republic*, Socrates contrasts a life of office-holding, of ruling, with the life of a philosopher (*Republic* 520e–521b; see also 496d–e and *Letter VII* 328a) and, later—and this would be the most stark contrast— the life of the philosopher with that of the tyrant (*Republic* 576b). In all these cases we are talking about a way of living or a kind of life that is either worthy or unworthy, praiseworthy or blameworthy, desirable or to be avoided.

6. The life that is most worth living for man is thus a quasi-divine life, something close to the life of mind and wisdom that Socrates, again in the *Philebus*, suggests would be "the most divine of lives" (*Philebus* 33b). In the *Phaedrus*, Socrates speaks of "a life of the gods" (*Phaedrus* 247e), while Adeimantus in the *Republic* argues that impressionable young men might get the wrong idea about how to lead "the best life" and so might live unjustly in an attempt to procure for themselves a "reputation for justice" and, thus, "a godlike life" (*Republic* 365b).

7. See also the *Seventh Letter*, where Plato recounts how he had hoped that Dionysus might, like Dion, have developed "a longing for the noblest and best life," that is, for the philosophical life, and so established a "blissful and true life throughout all the land" (*Letter VII* 327b–d).

8. It is perhaps telling that in tales of the afterlife in *Phaedo* and *Republic* we hear of humans being reincarnated as animals, *zōa*, of various kinds (kites, hawks, swans, ants, bees, apes, nightingales, lions) but never as plants. Plants would thus seem to be alive but, unlike animals in the strict sense and unlike humans, they do not seem to have a *bios*.

9. See *Protagoras* 321e, *Gorgias* 486c, *Laws* 842c, 936c, and *Critias* 114e, where the word seems to refer to the various requirements for life in

general—wood, metals, animals, etc. The term *biodotēs* has a similar meaning as *bios* at *Laws* 921a and *Republic* 381d.

10. "In truth, the constitution of these creatures has prescribed periods of life [*tou bion*] for the species as a whole. And each individual creature [*to zōon*] likewise has a naturally predestined term of life [*ton bion phuetai*], apart from the accidents due to necessity. For from the very beginning the triangles of each creature are constructed with a capacity for lasting until a certain time, beyond which no one could ever continue to live [*bion . . . biōiē*]" (*Timaeus* 89b–c).

11. We find a similar emphasis on duration in the *Phaedo*, where Socrates suggests that philosophers who have practiced dying their entire lives will not be afraid to die because they are going "to the place where there is hope of attaining what they longed for all through life [*dia biou*]" (*Phaedo* 67e; see also *Alcibiades I* 122a, *Philebus* 21b, 39e, 63e, *Symposium* 183e, 203d, 216e, *Phaedo* 75d). This would be the goal not only of the individual citizen in the state but of the state itself. As the *Laws* puts it, just as the individual must direct all his efforts "throughout the whole of [his] life" (*dia pantos tou biou*) (*Laws* 770d) toward the attainment of "excellence of soul," so the lawgiver must do everything possible so that "the whole of the community constantly, so long as they live [*aei dia biou pantos*], use exactly the same language, so far as possible, about these matters, alike in their songs, their tales, and their discourses" (*Laws* 664a; see 746a).

12. While *zōē* does not usually seem to suggest duration in this way, this passage from the *Laws* seems to be an exception to that general rule. The Athenian says he has been "considering by what means and by what modes of living [*ton bion*] we shall best navigate our barque of life through this voyage of existence [*dia tou plou toutou tēs zōēs*]" (*Laws* 803b).

13. In the *Timaeus*, the terms *bios* and *zōē* are used in the same phrase to describe those who, after living a life of ignorance—that is, "after passing a lame existence in life [*chōlēn tou biou diaporeutheis zōēn*]"—return again "unperfected and unreasoning to Hades" (*Timaeus* 44c). Finally, the Athenian in the *Laws* argues that women should "share with men in the whole of their mode of life [*tēs zōēs pasēs*]," the term *zōē*, rather than *bios*, being used, somewhat surprisingly, to describe this way of life (*Laws* 805d).

14. Liddell and Scott tell us that, in early writers, the present and imperfect of *bioō* were often simply supplied by *zaō*.

15. Socrates in the *Republic* speaks of the guardians "living happily" (*eudaimonōs biōsesthai*) and then of "the life they have lived [*tōi biōi tōi bebiōmenōi*] [being crowned] with a consonant destiny in that other world" (*Republic* 498c; see *Apology* 40d–e, *Gorgias* 485d, *Phaedo* 95c, 113e–114a, *Symposium* 178c, *Republic* 404b, 425e, 496e, 619c).

16. Whereas Socrates in *Alcibiades II* declares that "the state or soul that is to live aright [*orthōs biōsesthai*]" must hold fast to the knowledge of what is best (*Alcibiades II* 146e), Callicles in the *Gorgias* declares, as if in direct response to Socrates, that "he who would live rightly [*ton orthōs biōsomenon*] should let his desires be as strong as possible" (*Gorgias* 491e–492a).

17. Near the beginning of the *Republic*, Cephalus complains to Socrates that when he gets together with other older men they recall past pleasures and "repine in the belief that the greatest things have been taken from them and that then they lived well and now it is no life at all [*tote men eu zōntes, nun de oude zōntes*]" (*Republic* 329a; see also 330e, 362a, 561a, and *Laws* 679a for similar uses of the verb *zaō*.)

18. Alcibiades speaks in *Alcibiades I* of "our way of life"—that is, "our way of living in our cities" (*hōsper hēmeis zōmen en tais polesin*) (*Alcibiades I* 125c).

19. The verb *zaō* can even be used to describe not just the living of an individual in a state but the living of a state itself. The Athenian says in the first book of the *Laws*: "These are the two fountains which gush out by nature's impulse; and whoever draws from them a due supply at the due place and time is blessed—be it a State [*polis*] or an individual [*idiōtēs*] or any kind of creature [*zōon hapan*]; but whosoever does so without understanding and out of due season will fare [*zōiē*] contrariwise" (*Laws* 636e).

20. In later dialogues, the verb *zēn*, used as a substantive, means not so much "living" but "making a living," much like *bios* in earlier dialogues (*Sophist* 224d, *Laws* 678a, 847a).

21. The same can be said for the verb *bioteuō*, as in the following: "And so it is with the follower of each of the gods; he lives [*zēi*], so far as he is able, honoring and imitating the god, so long as he is uncorrupted, and is living his first life [*tēn tēide prōtēn genesin bioteuēi*] on earth, and in that way he behaves and conducts himself toward his beloved and toward all others" (*Phaedrus* 252d).

22. In the *Republic*, Socrates asks whether "the just soul and the just man then will live well [*eu biōsetai*] and the unjust ill," and then, a moment later, he affirms that "he who lives well [*ho ge eu zōn*] is blessed and happy, and he who does not the contrary" (*Republic* 353e–354a). Later in the dialogue, Socrates uses the first of these verbs when he asserts that "the tyrant's life [*biōsetai*] will be least pleasurable and the king's most," and, just a line later, the second to ask "how much less pleasurably the tyrant lives [*zēi*] than the king" (*Republic* 587b; see 591c). Similarly, in the *Gorgias*, Socrates says that the question at hand is "what course of life is best [*hontina chrē tropon zēn*]," the life to which Callicles is inviting him or "this life of philosophy [*tonde ton bion en philosophiai*]." In the end, says Socrates, they must determine "which of them is the one we ought to live [*biōteon*]" (*Gorgias* 500d; see 512a–b).

23. Michel Foucault, *The Hermeneutics of the Subject, Lectures at the Collège de France, 1981–1982*, trans. Graham Burchell (New York: Picador, 2005), 72–79.

24. See a similar use of *zēn* at *Laws* 807c–d.

25. In a similar passage in the *Gorgias*, Socrates seems to contrast simply living, *zēn*, for a certain period of time with living the best life. He argues that with regard to "living any particular length of time [*to zēn hoposondē chronon*], this is surely a thing that any true man [*alēthōs andra*] should ignore, and not set his heart on mere life [*philopsychēteon*]; but having resigned all this to Heaven . . . he should then proceed to consider in what way he will best live out his allotted span of life [*chronon biōnai hōs arista bioïē*]" (*Gorgias* 512d–e).

26. It is, in fact, from *zōē* (and/or *zaō*) that we get the verb *zōgrō*, meaning to capture someone alive, to take a living prisoner; see *Laws* 868b and *Republic* 391b. As for the adjective *zōtikos*, it characterizes the lively desire for emission or ejaculation, as well as the "lively" life-giving character of marrow (or semen): "And the marrow, inasmuch as it is animate [*empsychos*] and has been granted an outlet, has endowed the part where its outlet lies with a love for generating by implanting therein a lively desire for emission [*tēs ekroēs zōtikēn epithumian empoiēsas*]" (*Timaeus* 91b). In the only other passage in Plato in which this adjective is used, it is said that injustice, rather than killing or destroying its possessor, actually quickens him or makes him more lively, and thus, the argument will go, subject to greater punishment (*Republic* 610e).

27. In the *Phaedo*, Socrates speaks of a man as "still living [*zōntas*]" despite the fact that his body is being constantly destroyed or undone (*Phaedo* 87d). In the *Laws*, however, just to give a potential counter-example, the Athenian says that if the murderer of a free man, after being beaten and scourged by the public executioner, is "still alive [*eanper bioï*]," then he shall be put to death (*Laws* 872b). In the *Statesman*, *zēn* is used to describe life in a democracy that is without restraint—that is, without laws—and *bioō* life in a democracy with laws. Contrasting the various forms of government, the Stranger says that "if they are all without restraint, life [*zēn*] is most desirable in a democracy; but if they are orderly, that is the worst state to live in [*bioteon*]" (*Statesman* 303b).

28. In the *Symposium*, for example, it is implied that we have a "duty towards one's parents whether alive or dead [*peri goneas kai zōntas kai teteleutēkotas*]" (*Symposium* 188c). In the *Laws*, the Athenian makes it clear that we must help the living to live virtuously since it is too late to help them once they are dead: "But to him who is dead [*teteleutēkoti*] no great help can be given; it was when he was alive [*zōnti*] that all his relatives should have helped him, so that

when living his life [*ezē te zōn*] might have been as just and holy as possible, and when dead [*teleutēsas*] he might be free during the life which follows this life [*egigneto ton meta ton enthade bion*] from the penalty for wickedness and sin" (*Laws* 959c). See also *Republic* 366b, 465e, 503a, 618e–619a.

29. Socrates argues in the *Gorgias* that if one really wanted to harm one's enemy then it would be best for him not to be punished for his wrong-doing and "never die [*mē apothaneitai*], but be deathless [*athanatos*] in his villainy, or failing that, live as long a time as may be in that condition [*hopōs hōs pleiston chronon biōsetai*]" (*Gorgias* 481b).

30. Socrates in *Alcibiades I* says explicitly that *zōē* and *thanatos* are opposites: "And life and courage are the extreme opposites of death and cowardice [*thanatōi te kai deiliai enantiōtaton zōē kai andreia*]?" (*Alcibiades I* 115d)

31. While the soul makes all *zōa live*, it is not clear that it makes all *zōa move* or gives all *zōa self-movement*. After a long description in the *Timaeus* of the constitution of plants, Timaeus concludes that the plant, while retaining "its own native motion," "lives [*zēi*] and is not other than a living creature [*oukh heteron zōou*]" but "remains stationary and rooted down owing to its being deprived of the power of self-movement" (*Timaeus* 77b–c).

32. It is worth noting here that this same verb is used in the *Statesman* (at 271b) to describe the way in which the dead come back to life during the Age of Kronos. And then there is this happy thought from *Epinomis*: "And our time must be a short one . . . swiftly old age is upon us, and must make any of us loth ever to come to life [*anabiōnai*] again, when one reckons over the life one has lived [*logisamenon ton bebiōmenon heautōi bion*]—unless one happens to be a bundle of childish notions" (*Epinomis* 973d–974a).

33. This opposition can help explain, if not explain away, a few appearances of *zōē* that seem to run counter to everything I have been arguing. In the *Laws*, for example, the Athenian speaks of those who throw away their weapons and flee from battle as having gained "a life that is shameful [*zōēn aischran*] by speed of foot, rather than by bravery a noble and blessed death [*kalon kai eudaimona thanaton*]" (*Laws* 944c). Given all the passages in Plato that speak of a noble or ignoble *bios*, the use of *zōē* is here rather surprising, even if it is not completely unprecedented. Its appearance might be explained, however, by the fact that what is being underscored here is the relationship—or opposition—between death, *thanatos*, and life, which thus requires *zōē*, rather than *bios*. A similar line of reasoning could be used to explain this passage from the *Laws* where the Athenian says, in the midst of an argument that good souls go to the afterlife to be with good souls, and bad with bad, that all this is in accordance with "the decree that . . . alike in life and in every shape of death [*en te zōēi kai en pasi thanatois*], thou both doest and sufferest what it is befitting that like should do towards like" (*Laws*

904e). The same reasoning could be used to explain why *zaō*, rather than *bioō*, is used in the following: "wherein if he obeys the magistrates, he shall live as a private person [*idiōtēs zētō*], but if not, he shall be put to death" (*Laws* 952c; see *Cratylus* 395d). Finally, it could also help explain this odd reference in the *Laws* to "living souls": "Thus let it be laid down by law respecting the nurture and training of living souls [*ta . . . peri trophēn te zōsēs psychēs kai paideian*],—which when gained make life livable [*biōton*], but when missed, unlivable" (*Laws* 874d).

34. There seems to be a distinction between being and living in the *Republic*. Socrates says near the beginning of his construction of the city in speech, "now the first and chief of our needs is the provision of food for existence and life [*tou einai te kai zēn heneka*]" (*Republic* 369d); it is as if existing, *einai*, and living, *zēn*, were related in some unspecified way without being exactly the same.

35. One of the results of this analysis seems to be that *Alcibiades I* is a *late* dialogue, or, more probably, an earlier dialogue that was later rewritten.

36. All souls, as immortal, would thus be equally alive, it seems, just as, in another passage from the *Phaedo*, they are said to be equally good: "According to this argument, then, if all souls are by nature equally souls, all souls of all living creatures [*pasai psychai pantōn zōōn*] will be equally good" (*Phaedo* 94a). The point seems to be that, as immortal, every soul has to be good, even though, as Socrates will go on to argue, the soul will take on, at least provisionally, some of the character and habits of the bodies it inhabits.

37. I suggested in Chapter 3 that Mika Ojakangas's recent work *On the Greek Origins of Biopolitics* could be read as supporting, in large measure, the reading of the *Statesman* advanced here, even though my aim has not been to demonstrate the presence of biopolitics in the *Statesman* or in Plato more generally. Something similar could be said about Agamben's distinction between *bios* and *zōē*. While I do not wish to get into the debate surrounding this distinction in Agamben, I do want to argue that a distinction between *bios* and *zōē* is what helped make possible the emergence of something like *life itself* in Plato. What remains to be seen, then, is whether there is some more intrinsic relationship between the possibility of biopolitics and/or a thinking of *zōē* as bare life in Plato and the invention of life itself.

38. In the following passage from the *Laws*, it is a certain form of *bios* that is said to be most worthy of the name *bios*. "As compared with the life [*biou*] that aims at a Pythian or Olympian victory and is wholly lacking in leisure for other tasks, that life we speak of—which most truly deserves the name of 'life' [*bios eirēmenos orthotata*]—is doubly (nay, far more than doubly) lacking in leisure, seeing that it is occupied with the care [*epimeleian*] of bodily and spiritual excellence in general. For there ought to be no other

secondary task to hinder the work of supplying the body with its proper exercise and nourishment, or the soul with learning and moral training" (807c–d).

39. As for the construction of this Cosmos, we read: "but He, in the first place, set all these in order, and then out of these He constructed this present universe, one single Living Creature containing within itself all living creatures both mortal and immortal [*zōon hen zōa echon hapanta en hautōi thnēta athanata te*]. And He Himself acts as the Constructor [*dēmiourgos*] of things divine, but the structure of the mortal things [*thnētōn*] He commanded His own engendered sons to execute." Hence, "they, imitating Him, on receiving the immortal principle of soul [*archēn psychēs athanaton*], framed around it a mortal body [*thnēton sōma*]" (*Timaeus* 69c).

CONCLUSION: LIFE ON THE LINE

1. Jacques Derrida, "Final Words," trans. Gila Walker, in *The Late Derrida*, ed. W. J. T. Mitchell and Arnold I. Davidson (Chicago: University of Chicago Press, 2007), 244.

2. Jacques Derrida, *Rogues: Two Essays on Reason*, trans. Pascale-Anne Brault and Michael Naas (Stanford: Stanford University Press, 2005), 109.

3. Derrida, *Of Grammatology*, trans. Gayatri Chakravorty Spivak (Baltimore: Johns Hopkins University Press, 1976), 143, 208.

4. As Derrida shows in all these works, a certain conception of life is always in solidarity with a philosophical thinking that is at once anthropo-centric and pro-death penalty (as well as anti-democratic).

5. "And if they shall master these [desires and emotions] they will live justly [*dikēi biōsointo*], but if they are mastered, unjustly. And he that has lived his appointed time [*chronon bious*] well shall return again to his abode in his native star, and shall gain a life that is blessed and congenial [*bion eudaimona kai synēthē hexoi*]" (*Timaeus* 42b; see *Laws* 899b).

6. In the *Euthydemus*, Dionysodorus is able to bring the claim that the gods are *zōa* to its logical (or illogical) conclusion by arguing that one should then be able to buy, sell, and even sacrifice the gods like other *zōa* (*Euthydemus* 302a, 302d–303a).

7. It is unlikely that this is an authentic dialogue by Plato, but it is instructive to see the author struggling to reconcile various Platonic doctrines. It is from fire, one of the five elements (fire, water, air, earth, ether), that we get the "divine race of stars," which are "endowed with the fairest body as also with the happiest and best soul," "for each of them is either imperishable and immortal [*anōlethron te kai athanaton*], and by all necessity wholly divine, or has a certain longevity sufficient for the life of each [*ē tina makraiōna bion echein hikanon hekastōi zōēs*], such that nothing

could ever require a longer one" (*Epinomis* 981e–982a). (This relationship between life and fire is an undercurrent in both *Timaeus* and *Epinomis*. In the *Laws*, the Athenian speaks of the remnants of the human race after a catastrophe as the embers, the *zōpyron*, from which the human race will be rekindled [*Laws* 677b].) Either immortal (and so endowed, it would seem, with everlasting life) or else very long-lived: the author hesitates to come down squarely on one side or the other. He will not hesitate, however, to attribute life to the stars: "for never will fairer or more commonly owned images be found among all mankind, none established in more eminent places, none more eminent in purity, majesty, and life altogether [*xympasēi zōēi diapheronta*], than in the way in which their existence is altogether fashioned" (*Epinomis* 984a). Finally, it is worth noting that the author of *Epinomis* attributes intelligence, *phronēsis*, or rather "intelligent life," to the heavenly bodies insofar as they move in an orderly fashion: "Now that which has motion in disorder we should regard as unintelligent, acting like the animal creatures about us for the most part; but that which has an orderly and heavenly progress must be taken as strong evidence of its intelligence. For in passing on and acting and being acted upon always in the same respects and manner it must provide sufficient evidence of its intelligent life [*tou phronimōs zēn*]" (*Epinomis* 982b).

8. David White makes a very compelling case for reading the *Statesman* as a whole, and the myth of the two ages in particular, in relation to the *Philebus*. White starts from "the conviction that [the *Statesman*'s] imaginative myth of the reversed cosmos is indispensable to the teaching of the dialogue and that this teaching is primarily aporetic" (David White, *Myth, Metaphysics, and Dialectic in Plato's "Statesman"* [Burlington, Vt.: Ashgate Publishing Company], 2007, vii; hereafter abbreviated DW). In other words, the dialogue, as aporetic, "intentionally produc[es] blockages in thinking," thereby "inviting members of the Academy as well as modern readers to investigate statecraft in more philosophically appropriate directions" (DW 1). This, according to White, requires a rethinking of the status of the Good in other dialogues, and particularly the *Philebus*, which finally "connects measure with Forms through the essential connection between proportion and truth as elements of the Good" (DW 168). Hence "the *Philebus* resolves so many of the aporiae structuring the *Statesman*" (DW 191). (White cites Maurice Vanhoutte's *La Méthode Ontologique de Platon* [Louvain: Publications Universitaires de Louvain, 1956], 128–137 as precedent for his approach.)

This emphasis on the Good helps recast the myth of the two ages as a myth about the production of all beings in relationship to the mean and the mean in relationship to the Good: "According to the myth, all things in the cosmos are produced by the demiurge. But with the withdrawal of the demi-urge, things are threatened with complete dissolution. The demiurge returns, restoring and situating human beings in a mode of existence midway between the extremes of completely worldly satisfaction and virtual dissolution. Thus the 'human condition' may be characterized as a mean between extremes. But if what applies to human beings applies throughout the cosmos and if all beings in the cosmos are restored according to these specifications, then the existence of *all* beings is a mean between extremes, each type of being reflecting this mean in its own way" (DW 95). It's an ingenious argument that has many implications for the argument regarding life that I am trying to make here. I agree, for example, with White's claim that the myth of the two ages really is the *pivot* of the dialogue, the place where opposing motions are woven together—including life and death (DW 48).

9. We saw a similar borrowing in Chapter 6, where the best form of rule, the ideal form identified with the Age of Kronos, was the one in which the statesman makes "his science his law" (*tēn technēn nomon*) (297a)—that is, his law before and beyond the law, his law without writing and even without words (like "writing in the soul"). And we saw it when the Demiurge was called, using terms from the Age of Zeus, everything from "helmsman" to "demiurge" to "king," even "shepherd."

10. In the Age of Zeus, within this time that is the moving image of eternity, Zeus might indeed be understood as the god of life. It is perhaps not in complete jest, then, that, in the *Cratylus*, the etymology of Zeus's name is explained in terms of life. Socrates argues: "The name of Zeus is exactly like a sentence; we divide it into two parts, and some of us use one part, others the other; for some call him Zena, and others Dia; but the two in combination express the nature of the god . . . For certainly no one is so much the author [*aitios*] of life (*zēn*) for us and all others as the ruler and king of all. Thus this god is correctly named, through whom (*di' hon*) all living beings [*pasi tois zōsin*] have the gift of life (*zēn*)" (*Cratylus* 396a–b). But then there is this etymology of Kronos's name, which perhaps suggests the first reading outlined above: "It might seem, at first hearing, irreverent to call him the son of Kronos and reasonable to say that Zeus is the offspring of some great intellect [*dianoias*]; and so he is, for *koros* (for *Kronos*) signifies not child, but the purity [*katharon*], and unblemished nature of his mind [*nou*]" (*Cratylus* 396b).

11. It may well be that, as Miller argues, the *Statesman*, as a late dialogue, probably written when Plato was sixty-five or older, is also speaking about Plato's own absence, so that it is not just Socrates but Plato himself who is "testing for heirs" (Mitchell Miller, *The Philosopher in Plato's "Statesman"*, together with "Dialectical Education and Unwritten Teachings in Plato's *Statesman*" [Las Vegas: Parmenides Publishing, 2004], 116–117).

12. Rosen argues similarly: "The one thinker in the entire tradition of Western philosophy who is closest to what is today regarded as the rejection of Platonism is Plato himself" (*Plato's "Statesman": The Web of Politics* [New Haven: Yale University Press, 1995]; hereafter abbreviated SR).

13. According to Rosen, "weaving is the analogue to what the myth presents as rejuvenation through counternormal motion" (SR 102).

14. Don DeLillo, *Zero K* (New York: Scribner, 2016), 3.

Lightning Source UK Ltd.
Milton Keynes UK
UKHW040641271018

331299UK00001B/179/P